P9-AOR-947

The Future of the Defence Industries in Central and Eastern Europe

The Future of the Defence Industries in Central and Eastern Europe

SIPRI Research Report No. 7

Edited by
Ian Anthony

OXFORD UNIVERSITY PRESS
1994

Oxford University Press, Walton Street, Oxford OX2 6DP
Oxford New York Toronto
Delhi Bombay Calcutta Madras Karachi
Kuala Lumpur Singapore Hong Kong Tokyo
Nairobi Dar es Salaam Cape Town
Melbourne Auckland Madrid
and associated companies in
Berlin Ibadan

Oxford is a trade mark of Oxford University Press

Published in the United States
by Oxford University Press Inc., New York

© SIPRI 1994

British Library Cataloguing in Publication Data
Data available

Library of Congress Cataloguing in Publication Data
The Future of the Defence Industries in Central and Eastern Europe / edited by Ian
Anthony—(SIPRI research report : no. 7) Includes Index.
I. Economic conversion—Europe. Eastern—Congresses. 2. Defense
industries—Europe. Eastern—Congresses. 3. Europe. Eastern —Economic
policy—1989—Congresses. I. Anthony, Ian. II. Series.
HC244.Z9D424 1994 338.943—dc20 94–4993
ISBN 0–19–829184–1
ISBN 0–19–829189–2 (pbk.)

Typeset and originated by Stockholm International Peace Research Institute
Printed in Great Britain
on acid-free paper by
Biddles Ltd, Guildford and King's Lynn

Contents

Acknowledgements   viii
Preface   ix
Acronyms   x

1. Introduction   1
Ian Anthony
    I.   Background   1
    II.   Reasons for conducting this study   5
    III. The geographical scope of the study   10
    IV. Defining the defence industry   12
    V.   The structure of the report   14

2. Military doctrines in transition   16
Shannon Kile
    I.   Introduction   16
    II.   National threat perceptions and conflict scenarios   19
    III. A survey of selected military doctrines   25
    IV. Changes in force structures and levels   30
    V.   Concluding remarks   36
Table 2.1.   Former Warsaw Pact CFE Treaty ceilings   33
Table 2.2.   CFE-1A manpower limitations   36

3. Military expenditure in transition   37
Evamaria Loose-Weintraub and Ian Anthony
    I.   Introduction   37
    II.   Military expenditure data   38
    III. Selected country analyses   40
    IV. The dispersal of funds to industry   54
    V.   Concluding remarks   57
Table 3.1.   Estimates of aggregate military expenditure by Russia, 1991–93   41
Table 3.2.   Distribution of the Soviet/Russian defence budget, 1991–93   43
Table 3.3.   Reductions in procurement by the Russian Ministry of Defence in 1992 compared with the Soviet Ministry of Defence in 1991   44
Table 3.4.   Unit production by category and percentage changes in the former Soviet Union and Russia, 1990–92   45
Table 3.5.   Czechoslavakia's military expenditure allocation,1989–92   49
Table 3.6.   Polish Ministry of Defence expenditure allocation, 1989–93   51
Table 3.7.   Hungarian military expenditure allocation, 1989–93   52

Table 3.8.   Bulgarian military expenditure allocation, 1989–93   53
Table 3.9.   Romanian military expenditure allocation, 1990–93   55

4. Restructuring of the defence industry   58
Ian Anthony
        I.   Introduction   58
        II.   Intra-government changes   60
        III.   Changes in government–industry relations   69
        IV.   Changes in intra-enterprise relations   82
Figure 4.1.   Actors influencing defence industrial decisions in Russia, 1994   62
Table 4.1.   Distribution of production of fighter aircraft designed by Mikoyan and Sukhoi, 1993   86
Table 4.2.   Employment structure of defence industry enterprises, 1993   87

5. International dimensions of industrial restructuring   91
Ian Anthony
        I.   Introduction   91
        II.   Internationalization as a form of diversification   93
        III.   Teaming with foreign suppliers as a means of enhancing military capabilities   99
        IV.   Restoring defence industrial ties between Central and East European countries   103
Table 5.1.   Selected joint ventures with Russian aircraft industry, 1991–93   96
Table 5.2.   Government-to-government framework agreements for defence industrial co-operation, 1991–93   105

6. Arms exports   107
Ian Anthony
        I.   Introduction   107
        II.   Official data on the value of the arms trade   108
        III.   The volume of exports from Central and Eastern Europe   114
        IV.   Concluding remarks   121
Table 6.1.   Arms exports by Czechoslovakia, 1987–91   113
Table 6.2.   Arms exports by Poland, 1987–92   114
Table 6.3.   Arms exports by Hungary, 1971–89   114
Table 6.4.   Exports reported in 1993 to the UN Register of Conventional Arms for arms transfers in 1992 by Central and East European countries   115
Table 6.5.   Imports reported in 1993 to the UN Register of Conventional Arms for arms transfers in 1992 by Central and East European countries   116

Table 6.6.   Regional distribution of deliveries of arms and military
                equipment by the former Soviet Union in 1991   117
Table 6.7.   Distribution by weapon category of deliveries of arms
                and military equipment by the former Soviet Union
                in 1991   118

7. Conclusions   123
Ian Anthony
        I.   The defence industries of Central and Eastern Europe
                in context   123
       II.   Structural adjustment   125
      III.   Long-term developments   132

Appendix. List of participants in the 1993 SIPRI workshop   135
Index   137

Acknowledgements

In addition to the research team, the efforts of several SIPRI staff members were central to the production of this report. Project Secretary Cynthia Loo organized the workshop from which much of the information contained in this report was derived. The SIPRI librarians did their usual splendid job in tracking down elusive materials. Editor Don Odom prepared the book for publication.

Finally, I would like to thank SIPRI Director Adam Daniel Rotfeld for his advice and support throughout the process of researching and writing this publication.

Dr Ian Anthony
Arms Production and Arms Transfers Project Leader
June 1994

Preface

The main purpose of this SIPRI research report is to raise the level of understanding about what the industries of Central and Eastern Europe are and are not doing by going directly to the source of primary data— representatives of government and industry in these countries. With the financial support of the Swedish Ministry of Foreign Affairs, the SIPRI Arms Production and Arms Transfers Project began to assess the impact of the revolution which occurred in Central and Eastern Europe after 1989 on the defence industries of countries in this part of Europe.

Under any plausible scenario it will be many years before countries in Central and Eastern Europe establish a stable pattern of foreign relations, military doctrines, force structures, budget processes, a body of law governing economic activity and the procedures required to implement the law. Until these things are accomplished, the size and structure of the defence industry in the region will remain uncertain.

This report does not offer any blueprint for solving the web of complex and urgent problems resulting from attempts to restructure during a time of rapid political change. We hope that by providing a regular outlet for information and an analysis which places this information in a wider context SIPRI can make a modest contribution to the efforts of those people in the countries of Central and Eastern Europe who are responsible for finding solutions. This report, however, is by no means the final word on the subject from SIPRI. Indeed, a follow-on study is already in the project definition stage.

Dr Adam Daniel Rotfeld
Director of SIPRI
June 1994

Acronyms

ATTU	Atlantic-to-the-Urals (zone)
CFE	Conventional Armed Forces in Europe
CIS	Commonwealth of Independent States
CMEA	Council for Mutual Economic Assistance
COCOM	Co-ordinating Committee on Multilateral Export Controls
CSBM	Confidence- and security-building measure
CSCE	Conference on Security and Co-operation in Europe
GDP	gross domestic product
GOCO	government-owned, contractor-operated
ICBM	intercontinental ballistic missile
IFF	Identification Friend or Foe
IMF	International Monetary Fund
MTCR	Missile Technology Control Regime
NACC	North Atlantic Cooperation Council
NATO	North Atlantic Treaty Organization
NSWP	non-Soviet Warsaw Pact
OECD	Organisation for Economic Co-operation and Development
O&M	Operations and maintenance
SAM	surface-to-air missile
SLBM	submarine-launched ballistic missile
WTO	Warsaw Treaty Organization

Conventions used in the tables

$	US $, unless otherwise indicated
b.	Billion (thousand million)
m.	Million
..	Data not available or not applicable
–	Nil or a negligible figure
0	Less than 0.5

1. Introduction

Ian Anthony

East Europeans have little, if any, guidance as to how they should go about
the difficult business of defining a security policy that is politically feasible,
militarily credible and financially sustainable.[1]

I. Background

The collapse of the Warsaw Treaty Organization (WTO) and the
emergence of a new European security environment has left the
defence industries of Central and Eastern Europe with a production
over-capacity. Developed to meet the demands of armed forces whose
planning was against the contingency of a high intensity war, the new
strategic, economic and political environment has reduced the need
for perpetual force modernization. The terms of the 1990 Treaty on
Conventional Armed Forces in Europe (CFE Treaty) has established
binding commitments to reduce the level of specific types of equipment present in the inventory of regional armed forces, further
depressing the demand for major systems.

The defence industries of Central and Eastern Europe are trying to
cope with the crisis brought about by losing their most important
markets. Crisis strategies initially included seeking new export markets and maintaining production without sales to sustain employment.
In 1993 it became clear that the hoped-for boom in export sales is
unlikely to materialize while the levels of production which continued
into 1992 also appear to have stopped. Therefore, these short-term
strategies are no longer sustainable.

Existing production capacities could not be sustained by foreign
sales even in a buoyant market. In fact, the scale of the global arms
trade has been declining since 1987.[2] Continuing production where

---

[1] Gasteyger, C., 'The remaking of Eastern Europe's security', *Survival*, vol. 33, no. 2
(Mar./Apr. 1991).

[2] This finding is common to all agencies and institutes which estimate the volume of the
arms trade. For the most recent data, see SIPRI, *SIPRI Yearbook 1994* (Oxford University
Press: Oxford, 1994, forthcoming); Grimmett, R., *Conventional Arms Transfers to the Third*

there are no customers or in circumstances where the customer does not pay for goods ordered has contributed to the deterioration of financial systems in Central and Eastern Europe by adding to government deficits and the heavy burden of bad debt owed by industry. Producing systems for which there is no demand wastes resources, consumes capital and postpones adjustments which must at some point be made. In these conditions Julian Cooper has observed that 'a rapid and brutal downsizing of the defence industry' is both inevitable and imminent across the region.[3]

Nevertheless, the defence industries of Central and Eastern Europe represent a significant proportion of total global defence industrial capacity. Russia possesses the largest defence industry in the world (albeit one in deep crisis) and even after a major rationalization the industry will be by far the largest in Europe. Ukraine also has a significant defence industry (although orders of magnitude smaller than that of Russia). Moreover, Ukraine has inherited important research and design capacities in the area of transport aircraft and tanks from the former Soviet Union. The structure of Ukrainian industry is unique, with control over the production cycle for intercontinental ballistic missiles but no capacity to manufacture an assault rifle or a light vehicle.

Compared with the former Soviet Union, the Czech Republic, Poland, Romania and Slovakia all have small defence industries which produce a limited range of products. This is even more the case for Bulgaria and Hungary. Compared with other European countries, however, their defence industrial capacities are significant in terms of the volume of equipment they could produce for customers with economic means to buy them.

Although the data on the extent of indigenous military research and development (R&D) in Central European countries are poor it seems clear that only Russia has the human and material resources to design and develop a range of new major combat systems. Judit Kiss has expressed this as follows:

World 1984–92 (Congressional Research Service: Washington, DC, June 1993); Arms Control and Disarmament Agency, *World Military Expenditures and Arms Transfers* (US Government Printing Office: Washington, DC, 1990).

[3] Julian Cooper, speaking at the workshop on Conventional Arms Proliferation in the 1990s, Carnegie Endowment for International Peace, 9–10 July 1992.

Local military R&D was authorized by the highest WTO leadership—which in practice meant the Soviet authorities. Bulgaria specialized in radio and computer production; telephone and switchboard equipment was made by the former GDR and Hungary carried out research in telecommunications. This meant that at least part of the production was based on local R&D though local experts dispute its extent. Some sources deny independent research but confirm local technology development.[4]

The range of products manufactured by the defence industries of countries other than Russia is likely to be limited to the incremental modification of existing designs. At the point where they are ready to move to a new generation of equipment—a point which is some way in the future—it is unlikely that the smaller Central and East European countries will be able to meet their requirements by developing new products themselves. They are more likely to buy finished systems or production licences from abroad.

In Russia it now appears that very few new major weapon programmes are being initiated while several large projects have been suspended at the final development stage. It is true that some major systems with new and previously unknown designations have been advertised at various international arms fairs. However, in all cases these have turned out to be modified versions of existing designs.

In present conditions design bureaus are unable to offer competitive salaries to designers. Whereas the liberalization of some prices increased the costs of industry, the production prices for goods ordered by the Russian Government are limited by administrative decision. As a result the defence industry has not been able to generate revenue (and thereby raise salaries) by increasing prices as other producers have.[5] Designers are also frustrated by their current under-employment—something about which the producers on which they rely cannot do very much. Young designers in particular are leaving or trying to leave although few if any have been made redundant, reflecting the desire to hold these skilled teams together.[6] This has threatened the capacity of Russia to develop competitive major sys-

---

[4] Judit Kiss at the SIPRI workshop on The Future of the Defence Industries of Central and Eastern Europe, 29–30 Apr. 1993. For the list of workshop participants, see the appendix.

[5] Nina Oding, Head of Research, Leontief Institute, St Petersburg, speaking at the SIPRI workshop on The Future of the Defence Industries of Central and Eastern Europe, 29–30 Apr. 1993.

[6] Oleg Samoylovich, Moscow Aviation Institute, speaking at the SIPRI workshop on The Future of the Defence Industries of Central and Eastern Europe, 29–30 Apr. 1993.

tems independently. It may be that Russia will depend in future on foreign financing: China is perhaps the most likely partner for new programmes.

Several of the arms industries of the region—in particular in Czechoslovakia and Poland—were independent centres of innovation in military technology development before 1939. After 1945 they were reconstructed as a coherent entity in which the Soviet Union exercised both technical and political control over decisions taken by countries with systems based on state socialism.

The dissolution of the WTO had a severe impact on this integrated production system but it did not entirely eliminate it. Moreover, Central European countries are beginning to investigate restoring bilateral links in the technical and industrial sphere. In the present budget environment the need for practical co-operation may replace the political dimension as the driving force in bilateral relations throughout the region. As part of the former Soviet Union, Ukraine had a higher degree of interdependence of markets and production with Russia than other Central European countries. As a result, the need for continuity in industrial collaboration with Russia has been greater in Ukraine regardless of difficulties in their bilateral political relationship.

There are no uniform national approaches to dealing with the consequences of continued interdependence, and a wide range of different responses are predicted. Some believe a 're-nationalization' of military industrial policy to be likely. Under this scenario military expenditures will favour domestic producers, and specific equipment procurement choices will reflect existing production capabilities. Moreover, in time new industrial capabilities to meet important equipment requirements may be developed. Others argue that for practical reasons this will be impossible except on a piecemeal basis. National military R&D capabilities are inadequate and there is no prospect that this will change given the fall in overall military expenditures. Consequently, the only alternative is to collaborate across borders. Central and East European countries are seeking joint arrangements such as the Commonwealth of Independent States (CIS) and the Visegrad Group.[7] Neither arrangement seems likely to obviate

---

[7] Central Europe comprises the non-Soviet WTO countries. Eastern Europe includes the newly independent states which have emerged on the territory of the former Soviet Union. The Visegrad Group consists of the Czech Republic, Hungary, Poland and Slovakia. This

the need for domestic arms production as long as national armed forces remain the basis for national defence planning. In addition, several Central European countries would like to join the North Atlantic Treaty Organization (NATO), although this is unlikely to occur in the near term. In practice, a mix of self-reliance and foreign collaboration is not only possible but the most likely course of action.

II. Reasons for conducting this study

The belief that armament policy should be seen as an essential element of defence policy is common among states. Moreover, in shaping this policy the participation of the three sides of the 'Iron Triangle' identified by Gordon Adams is also common to all countries with significant defence industries.[8] The nature of arms procurement dictates a need for close liaison between the armed forces (which define the technical character of the market in which producers of military equipment must operate); the government (which defines the economic character of the market in which producers of military equipment must operate); and the industry (which produces and manufactures the goods). However, this brief introduction indicates that the structural changes under way in the defence industries of Central and Eastern Europe are not a limited adjustment to temporary reductions in funding. Fundamental changes have already occurred and further changes are inevitable.

The main purpose of this study is to investigate the nature and extent of these changes in the defence industries in Central and Eastern Europe. Understanding developments in Central and Eastern Europe requires direct contact with the sources of primary data—representatives of government and industry—in these countries. Comparative analysis based solely on the official aggregate data at present in the public domain is not possible because of problems associated with the data. Some of these problems are explored in greater detail in chapter 3 of this report.

This report is based on presentations made at the SIPRI workshop on The Future of the Defence Industries in Central and Eastern

---

report concentrates on the countries of the Visegrad Group, Romania, the Russian Federation and Ukraine.

[8] Adams, G., *The Iron Triangle: The Politics of Defense Contracting* (Council on Economic Priorities: New York, 1981).

Europe, held in Stockholm on 29–30 April 1993. In selecting participants for the workshop, representatives of two groups were given
priority: the arms industry and government ministries responsible for
making defence industrial policy.[9]

This report is not a volume of proceedings nor a synthesis of views
presented at the workshop but an effort to identify the factors
considered decisive in shaping the future scale and structure of the
arms industry which draws on the data and ideas presented at the
meeting. Nevertheless, full responsibility for the views expressed
rests with the author of each chapter, and nothing published in this
report necessarily reflects the views of the governments or enterprises
which workshop participants (who were invited in their private
capacity) represented.

Widely divergent views were expressed at the workshop, and no
consensus was sought. Two basic 'fault lines' emerged in the
meeting. The first stemmed from the different views which industry
and defence ministries naturally hold on defence industrial issues.
These differences are not unique to Central and Eastern Europe but
exist the world over. They are a natural by-product of the different
tasks and responsibilities of government and industry. The old
mechanisms for managing these differences of interest have broken
down in Central and Eastern Europe, and new ones have not yet
emerged.

The second area of divergence identified at the meeting stemmed
from the unique economic and political transformation which is
taking place in Central and Eastern Europe. The defence industries of
these countries have not been insulated from the effects of a simultaneous transition from single-party systems to more representative
forms of government and from planned to market economies.

Any study of the defence industries in Central and Eastern Europe
is at least as much a study of political and strategic developments as it
is a study of technical and economic developments. In fact, under
present conditions even the distinction between politics and economics is probably of limited use. Political change has been a precondition for economic reform while attitudes towards economic
reform have become an important means of defining political align-
ment.

---

[9] A list of workshop participants is contained in the appendix.

Sergey Kortunov, Head of the Directorate of Export Control and Conversion in the Russian Ministry of Foreign Affairs, has observed that 'letting market forces consolidate the [defence] industry cannot be done in Russia. Such a significant part of national resources cannot be let just skip away'.[10] However, recent analyses of economic developments tend to emphasize measures other than conversion which would have great indirect consequences for the defence industry.

These include macroeconomic measures—including an end to all financial support for state-owned enterprises beyond purchasing their products. Anders Åslund suggests 'enterprises only adjust when they are compelled to do so. Thus, until the money crunch is allowed to hit the enterprises, neither macroeconomic nor microeconomic improvements are likely'.[11] However, during a period when the executive branch of government favoured this approach it proved impossible to implement in Russia. President Boris Yeltsin has continued to allow his Minister of Finance to make supplementary allocations to the defence industry while the Minister of Defence continues to place orders with industry that the government has no means to support. By the end of 1993 some economists had concluded that the idea of cutting subsidies was 'a commitment without any credibility which is eroding the reputation of the government'.[12]

Elsewhere in Central Europe, governments have also arranged financial transfers to industry in spite of the lack of demand for its products. These producers now owe enormous debts to state-owned financial institutions and to each other.

Reducing the size and influence of the defence sector—effectively a 'state within the state' under the previous system—and making it accountable to a representative civil authority are important political objectives within the overall process of reform. In Russia especially, civil authorities have not found an effective means by which to 'bell the cat'—political leaders must court military support if their policies are to succeed. The military wields a disproportionate and perhaps

[10] Kortunov, S., 'Conversion in Russia: the search for the most efficient option', paper presented at the NATO–CEE Conversion Seminar, Brussels, 20–22 May 1992.
[11] Åslund, A., *Systemic Change and Stabilization in Russia* (Royal Institute of International Affairs: London, 1993), p. 17. The same line is developed by authors who include Boris Fedorov and Anatoliy Chubais in Åslund, A. and Layard, R. (eds), *Changing the Economic System in Russia* (Pinter: London, 1993).
[12] Sapir, J., 'The Russian defence related industries conversion process', Centre d'Études des Modes d'Industrialisation, Paris, Oct. 1993, p. 34 (mimeographed).

growing political influence because it is a powerful ally for politicians unable to establish their authority over one another.

One strand in recent analysis of economic policy ignores the defence industry *per se* and focuses on measures required in order to grow the civil sectors of the economy. This would have a major impact on the defence industry by providing employment (removing a powerful political motivation for sustaining investment in the arms industry) and an alternative means of sustaining basic research and the national technology base.

There seems to be growing agreement that the civil industrial sector can best be developed by breaking up large production units and encouraging the growth of small businesses. Some recommend that these should operate autonomously and perhaps under private ownership and management. Reorganizing large multi-product industrial associations into smaller, more product-specific entities is a pattern of industrial restructuring in the Czech Republic, Hungary, Poland and Slovakia, and appears to be occurring in important arms-producing regions in Russia such as St Petersburg and Nizhniy Novgorod.

In Russia this process is intended to concentrate national military industrial capacities in a small number of large industrial business units. These 'prime contractors' continue to be required for programme management—to integrate and co-ordinate the activities of lower-tier suppliers. In the countries of Central Europe government and, especially, the defence manufacturers currently responsible for systems integration themselves seem to have concluded that they are unlikely to be able to retain current capabilities. Industrial restructuring is therefore an effort to diversify or, preferably, move away from defence-related production. In future the companies which are now being created expect to derive a significant percentage (if not the majority) of their sales by supplying sub-systems or components to overseas prime contractors, preferably in the civil sector.

Whether or not these plans for industrial restructuring can be implemented is very much an open question in present circumstances. Although a dramatic shift is certainly taking place in the defence industries across Central and Eastern Europe there is little sign that it is in line with any industrial policy. Rather, it is a market-led response to the collapse in demand for the products made in this manufacturing sector. As a result several of the countries in the subregion are maintaining levels of production which ideally they would not wish to.

To the extent that countries in Central and Eastern Europe intend to
retain national industrial capabilities to support their armed forces,
government intervention will be needed. The natural tendency for
commercial companies will be to shed any capacity which is not
being utilized. The decree by President Yeltsin of July 1992 requiring
industrial enterprises to maintain a mobilization potential drew an
exasperated response from Vladimir Alferov, executive director of the
League of Assistance for Defence Enterprises: 'all the directives from
above are driving our directors mad: increase the output of civilian
goods but do not decimate military production. What on earth does
this mean?'[13]

A different form of government intervention—to transform specific
industrial capacities—is increasingly being avoided. The failure of
conversion reflects the fact that political directives are usually drawn
up without an adequate understanding of arms production or the wider
manufacturing process. Programmes such as those made by Gosplan
(the Soviet State Planning Committee) between 1988 and February
1990 demanded increases in the volume of civilian production by
military production facilities, focusing on product types perceived to
be in short supply.[14] The models of conversion applied after 1988
were both acknowledged to have failed by 1993.[15]

Establishing a commercial production line in a defence plant
generated civil products but paid no attention to the differences in
marketing and distributing them compared with military equipment—
generating anecdotes about warehouses brimming with unsold

[13] *Kommersant*, no. 15 (1993).

[14] Conversion efforts are described in Cooper, J., *The Soviet Defence Industry: Conversion
and Reform* (Pinter: London, 1991); Twygg, J., 'Order out of chaos: conversion of the Soviet
defense industries', paper for the Council on Economic Priorities project on US–Soviet
Military Expenditure, Jan. 1991; Kortunov, S., 'Conversion in Russia: the search for the most
efficient option', paper presented at the NATO–Central and Eastern Europe Conversion
Seminar, Brussels, 20–22 May 1992; Checinski, M., *Military–Economic Implications of
Conversion of the Post-Soviet Arms Industry*, Research Paper no. 75, Marjorie Mayrock
Center for Soviet and East European Research (Hebrew University of Jerusalem: Jerusalem,
winter 1992); Malleret, T., *Conversion of the Defense Industry in the Former Soviet Union*,
Institute for East–West Security Studies, Occasional Paper 23 (IEWSS: New York, 1992);
Izyumov, A., 'The Soviet Union: arms control and conversion—plan and reality', ed. H.
Wulf, SIPRI, *Arms Industry Limited* (Oxford University Press: Oxford, 1993); Khroutskiy,
V., Koulik, S. and Ushanov, Y., *Russian Military Industry: Present Realities and Conversion
Efforts* (Center for Conversion and Privatisation: Moscow, 1993).

[15] Cipriano, F., Statement before Joint Hearings of the Committee on Foreign Affairs and
Security; Committee on Budgets; Committee on Economic and Monetary Affairs and
Industrial Policy; and Committee on External Economic Relations, European Parliament,
26 Apr. 1993.

toasters and coffee pots. A second model consisted of efforts to shift the assets of a defence plant to civil production. This led to the development of goods unsuited to civilian use or too costly for civilian consumers—the famous 'titanium baby carriage' and the washing-machine with 50 wash-cycles. Even individuals not sympathetic to the military–industrial complex have found many of the criticisms levelled against conversion 'reasonable and legitimate'.[16]

In 1993 the Russian Supreme Soviet published a new programme for the conversion of the defence industry to cover the years 1993–95, identifying 14 priority areas for conversion but not explaining how the problems which led to the failure of past plans would be overcome.[17] The authors of the programme acknowledged that 'the substitution of the centralised management of the national economy by a mainly economic market regulation' was a precondition for successful conversion rather than conversion being a mechanism for achieving systemic change in the economy.

III. The geographical scope of the study

Looking at socialist countries in Europe *en bloc* was logical before the end of the cold war. However, as new difference emerge in the political and industrial systems of these countries, discussing their arms industries in the subregional context is likely to become more difficult.

The goals of defence planners in Central Europe are similar to those of governments in Western Europe. The question 'how much is enough?' has been reformulated as 'how little is sufficient?' but the issue is the same: how to provide the armed forces with equipment without imposing an intolerable burden on the economy. There is no technical formula for establishing sufficiency in defence. Moreover, there is no consensus across the region or in individual countries about what kinds of armed forces are required in the new political environment; what should be the level and sophistication of their equipment and what constitutes a tolerable economic burden.

[16] See Izyumov (note 14), p. 114.
[17] V. Y. Vitebskiy, Deputy Chairman, Committee on Industry and Energy, Supreme Soviet of the Russian Federation, *Programme on Conversion of the Defence Industry 1993–95*, Joint Hearings before the Committee on Foreign Affairs and Security; Committee on Budgets; Committee on Economic and Monetary Affairs and Industrial Policy; Committee on External Economic Relations, European Parliament, 26 Apr. 1993.

A divergence between the Russian experience and that of Central European countries is especially clear. Moreover, these are differences in kind and not in degree. Where it comes to threat perception and the probability of the use of military force the debate in Russia has a different dynamic from that in the other countries.

Governments have not yet decided what kind of defence industry they would like to preserve under optimum conditions. Russia has defined foreign and security policy goals and seems ready to enunciate a military doctrine to pursue those goals which requires a very significant defence industry. In early November 1993 President Yeltsin announced that a new draft doctrine for Russia had been prepared. While the document containing the draft of the doctrine was not published, many of the details were released to the press. These, together with other statements by Defence Minister Pavel Grachev and First Deputy Defence Minister Andrey Kokoshin, suggest that Russia continues to use the United States as a yardstick in measuring the quantitative and qualitative elements of its military effort.[18]

The approach to national security includes the assumption that under certain circumstances Russian forces might initiate military actions in neighbouring countries. The belief that localized conflicts may escalate and draw in extra-regional military forces and that this requires Russia to preserve large offensive forces and levels of force sufficient to defeat several combined adversaries are also threads running through the draft doctrine. Under these conditions the decisions taken by Russia about its foreign and security policy are likely to have a significant impact on perceptions—and choices about resource allocation for defence—made by all of Russia's neighbours but particularly the countries of the former Soviet Union.

In the other Central and East European countries the situation is very different from that of Russia in that doctrines are limited to operations on national territory and not outside the borders of the country. The contingencies which are of most concern to the military are coping with the possible outbreak of civil war and the collateral impact of a war fought in or between neighbouring countries.[19]

There is a consensus in industry that the articulation of a new doctrine is a precondition for medium-term planning. While there is no

---

[18] The evolution of Russian military doctrine is summarized in Dick, C. J., 'Initial thoughts on Russia's draft military doctrine', *British Army Review*, vol. 104 (1993), p. 63.

[19] de Weydenthahl, J. B., 'Poland's security policy', Radio Free Europe/Radio Liberty, *RFE/RL Research Report*, vol. 2, no. 14 (2 Apr. 1993), p. 31.

intention among arms producers of Central and Eastern Europe to resign from export sales, orders from domestic ministries of defence are central to their future viability. However, few governments have come far in defining what will be required of their armed forces in operational terms.[20]

IV. Defining the defence industry

As noted above, differences emerging between the countries of Central and Eastern Europe are making it difficult to make any general statements about this subregion. Equally, it is difficult to isolate the defence industry as a discrete entity. The producers involved in providing the range of manufactured goods used by the military are far from being a homogeneous industrial sector.[21]

The arms industry can be more closely defined. It is confined to those entities involved in the development, production and distribution of lethal items and their delivery systems. In its 1991 report on the defence industrial base, the US Department of Defense concentrated on companies with 'primary responsibility for developing, designing and producing weapon systems'.[22] This definition would include the suppliers of components and sub-systems to the large industrial producers responsible for major platforms. Nevertheless, this would exclude a significant fraction of defence-dependent industrial activity. An effective military force depends on combat support equipment of various kinds and items that have civilian as well as military uses. In the USA each military service, the Defense Logistics Agency and the Office of the Secretary of Defense also maintain offices collecting industrial data and assessing the impact of government decisions on a wide range of industries.

Organizing the discussion around the customer presents different problems since the armed forces and civilians employed in defence consume many products ranging from tents to telephones, bootlaces to stationery. If the analysis is confined to the economic impact of changes in defence spending, the issue is simpler. However, the

[20] Elements of this debate are examined in chapter 2.

[21] The problems of definition are introduced in Taylor, T. and Hayward, K., *The British Defence Industrial Base: Issues and Options* (Brassey's: London, 1989).

[22] *The DOD Defense Industrial Base*, Joint Report to Congress by the Undersecretary of Defense (Acquisition) and the Assistant Secretary of Defense (Production and Logistics), Nov. 1991.

impact of changes in resource allocation on political and strategic questions requires a definition which isolates the capacity to produce goods and services needed to perform operational requirements.

In 1986 the British House of Commons Defence Committee defined the defence industrial base as 'industrial assets which provide key elements of military power'. The phrase 'key elements' cannot be defined precisely but this nevertheless seems the most useful definition.

For the countries of Central and Eastern Europe some further complications exist in what is already a conceptually problematic area. First, the literature on theoretical aspects of the defence industry is overwhelmingly related to NATO countries. Second, there is no single word with which to describe the individual units which make up the defence industry in the Central and East European countries. These are not companies in the accepted sense of that word because a company exists within a legal system which regulates its economic transactions, ownership and obligations to shareholders, employees and consumers. No such system has yet been put in place in Central and Eastern Europe. In many cases the entities engaged in production are not single factories or workshops (although such plants do exist) but associations linking the activities of more than one facility located at more than one site. For want of a better alternative these units are called 'enterprises'.

In addition to differences in the type of actor there are two other types of diversity within the defence industry: those stemming from the position of an enterprise in the production hierarchy and those stemming from the nature of the product being made.

The industry is organized in a vertical hierarchy which consists of at least six levels.[23] The levels are (from the top down): integrated weapon and information systems; major weapon platforms; complete weapons and communication kits; sub-systems; components; and materials. Of the countries of Central and Eastern Europe, only Russia is represented at each level. From an industrial perspective there seem to be no enterprises active at all levels, which creates interdependence

[23] Walker, W., Graham, M. and Harbor, B., 'From components to integrated systems: technological diversity and integrations between the military and civilian sectors', eds P. Gummett and J. Reppy, *The Relations between Defence and Civil Technologies* (Kluwer Academic Press: Dordrecht, 1988). There are other models of the industry but that in Walker, Graham and Harbor is the most useful.

between manufacturers of different kinds and between manufacturers and material suppliers.

The importance of non-lethal products to effective military operations has created an industry consisting of 'a few sectors which are defence unique and a far larger number of sectors which are fully integrated in terms of the civilian and military portions of those sectors'.[24]

To place an enterprise within the defence industry it is necessary to have information about both its location in the product chain and the type of activity undertaken. The behaviour of individual units which make up the defence industry will differ greatly depending on this location. To offer two extreme examples, a manufacturer of major weapon platforms located in the defence-unique sector will have very different problems and possibilities compared with a component maker in an integrated military/civilian sector. The aggregate information about Central and Eastern Europe in this regard remains very limited, although it is growing.

Even if this information were accessible it would still be difficult to fit Central and Eastern Europe into the existing theoretical framework describing the behaviour of the defence industry because the system of regulation, financing and marketing required for similar products is very different. The fragmented anecdotal evidence available and the lack of applicable theory make describing the behaviour of enterprises in Central and East European countries rather like navigating through the snow. While some landmarks seem familiar, their appearance has been altered to the point where identities become uncertain.

V. The structure of the report

The chapters which follow are organized around the broad elements on which the future development of arms industries in Central and Eastern Europe will depend.

Chapter 2 summarizes the discussion on military doctrine and force structure which has taken place in Central and Eastern Europe. This cannot be a comprehensive treatment of the subject, although SIPRI is working on a more detailed study of the specific case of Russia—

---

[24] Gansler, J., 'Supplier restraint: towards conversion?', paper presented to the International Institute for Strategic Studies (IISS) Conference on Conventional Arms Proliferation, 5–7 May 1993.

where developments are likely to have a decisive impact on the behaviour of other European countries.[25]

Chapter 3 of the report examines the recent trends in the defence budget and military expenditure among Central and Eastern European countries of the region including a discussion of defence budget in the context of overall government spending and the allocation of resources within the defence budget.

Chapter 4 considers the steps taken to restructure the defence industry in Central and Eastern Europe thus far. Intra-governmental changes, changes in government–industry relations and changes in the linkages between industrial concerns form the elements of this discussion. However, the information available is too limited and fragmented to permit any in-depth treatment of these issues, on which a great deal more work is required.

Chapter 5 looks at international linkages of the defence industry both within the region and with extra-regional partners. The focus is on international linkages in the military area rather than efforts to use industrial joint ventures or foreign direct investment as a means of diversification away from defence.

Chapter 6 examines the current role of Central and East European countries in the international arms trade.

Chapter 7 reviews the findings of presented chapter 6 and suggests possible future patterns of development for the industry over the short, medium and long term.

[25] SIPRI is currently engaged in a multi-author three-year study assessing the future orientation of Russian foreign and security policy, Russia's Security Agenda. This project is led by Dr Vladimir Baranovsky, SIPRI Senior Researcher.

2. Military doctrines in transition

Shannon Kile

I. Introduction

The dissolution of the Warsaw Pact and the end of the cold war have fundamentally transformed the strategic landscape of Europe.[1] For the former Warsaw Pact member states, the focus of defence planning for nearly four decades—countering a putative threat from NATO—has disappeared, taking with it the central rationale for maintaining enormous cold war-era defence establishments. Against a background of difficult economic transitions and sorely squeezed defence budgets, the governments of these states now confront defence planning problems of an entirely different character: eliminating surplus military equipment, personnel and industrial production capacity; and adapting national armed forces to an unsettled security environment lacking an immediate or well-defined threat.

National responses to these challenges have varied, as states composing the once-cohesive military bloc go their own way on defence matters. Liberated from the constraints of an alliance framework in which Soviet security interests and concerns were paramount, governments across the region are in the process of elaborating new military doctrines designed to support independent foreign and security policies.

Among the former non-Soviet Warsaw Pact (NSWP) member states in Central and Eastern Europe, military planning has undergone a 'turn to the defensive' that explicitly repudiates the unambiguously offensive strategy enshrined in Warsaw Pact military doctrine. Defence ministries in the Czech Republic, Hungary, Poland and, to a lesser extent, Slovakia have embarked upon far-reaching and often difficult programmes of military restructuring and reform aimed at bringing national armed forces into line with new independent military doctrines oriented towards defensive operations conducted exclusively on national territory.

[1] In common Western usage, 'Warsaw Pact' describes any of the politico-military entities and activities of the Warsaw Treaty Organization (WTO).

The changes under way in their force postures are extensive, requiring considerably greater adjustments than the successive parings of military equipment and manpower mandated by conventional arms control agreements.[2] Re-configuring national armed forces for the defensive roles and missions called for by new military doctrines involves an overhaul of the way in which they have been trained, equipped and deployed for nearly 40 years. It also involves a comprehensive organizational restructuring of military command arrangements and a defence planning process long centred on the Soviet Ministry of Defence (MOD).

In Russia, the recasting of the doctrinal underpinnings of post-Soviet defence planning has been less extensive and involved a different civil–military policy-making dynamic from that in the former NSWP states in Central Europe. The process of elaborating a new military doctrine has been caught up in the broader political struggle within Russia over the purpose of the armed forces and the degree of policy-making autonomy to be retained by the professional military and its institutional allies.[3] With the waxing of the senior military leadership's political influence, some of the traditional themes of the former Soviet General Staff have reappeared in the military doctrine published in November 1993. Consistent with Moscow's increasingly assertive foreign policy tone, the new doctrine implies a substantial defence effort and large armed forces possessing relatively robust offensive capabilities.

In Ukraine, a new military doctrine won the approval of the Rada (Parliament) in October 1993 following the rejection of earlier drafts. Defence planning has been complicated by the suddenness of the former Soviet republic's political transition to independence. The process of building national armed forces began without the benefit of a

---

[2] The CFE Treaty, signed by the 22 NATO and WTO member states in Nov. 1990, imposes quantitative ceilings on national holdings of five categories of major combat equipment. The follow-on Concluding Act of the Negotiation on Personnel Strength of Conventional Armed Forces in Europe (CFE-1A Agreement), signed in July 1992, caps total national military manpower levels. For the text of the CFE Treaty, see SIPRI, *SIPRI Yearbook 1991: World Armaments and Disarmament* (Oxford University Press: Oxford, 1991), appendix 13A. For the text of the CFE-1A Agreement, see SIPRI, *SIPRI Yearbook 1993: World Armaments and Disarmament* (Oxford University Press: Oxford, 1993), appendix 12C.

[3] For an overview of the development of post-communist civil–military relations in Russia, see Arnett, R., 'Can civilians control the military?', *Orbis*, vol. 38, no. 1 (winter 1994), pp. 41–58. See also Taylor, B., 'Russian civil–military relations after the October uprising', *Survival*, vol. 36, no. 1 (spring 1994), pp. 3–29.

widely accepted domestic political understanding of Ukraine's national security interests, or even a coherent and well-developed security policy debate.[4] In the absence of a domestic security policy consensus, the drafting of the basic principles of a new military doctrine proved to be a potent well-spring of political acrimony within the parliament, particularly over the issue of the country's future nuclear weapon status.

The nature of military doctrine

The military doctrines taking shape in most of the states tied to the old Soviet bloc do not provide systematic guidelines for defining defence requirements or methodologies upon which decisions about force structuring, weapon development and force employment can be founded. Rather, they essentially offer political blueprints for military reform that reflect the general re-direction of the foreign and security policies of these states in the wake of the dissolution of the Warsaw Pact and the end of the cold war.

Official presentations of post-Warsaw Pact national doctrines, even by Central and East European governments keen to jettison all remnants of the Soviet military legacy, evince a different conceptual focus from that found in Western doctrinal discussions.[5] Greater emphasis is given to the foreign policy dimensions of doctrinal reforms than to the considerations of operational techniques and methodologies that tend to drive debates within NATO and the United States.

In this regard, national doctrines hark back to their Soviet antecedents.[6] The Soviet usage of the term 'military doctrine' had no exact equivalent in the Western lexicon. It consisted of two elements: an overarching military–political component which functioned as a virtual surrogate of what would be called national security policy in the West; and a subordinate military–technical component which

[4] See Mihalisko, K., 'Security issues in Ukraine and Belarus', ed. R. Cowen Karp, SIPRI, *Central and Eastern Europe: The Challenge of Transition* (Oxford University Press: Oxford, 1993), pp. 246–48.

[5] Lachowski, Z., 'The Second Vienna Seminar on Military Doctrine', *SIPRI Yearbook 1992: World Armaments and Disarmament* (Oxford University Press: Oxford, 1992), p. 499.

[6] The national military doctrines of the NSWP states, with the notable exception of Romania's, were subordinate to, and virtually indistinguishable from, Warsaw Pact doctrine. In turn, Warsaw Pact doctrine was the coalitional component of Soviet military doctrine and as such mirrored Soviet military thinking. Alexiev, A., Johnson, A. R. and Dean, R. W., *East European Military Establishments: The Warsaw Pact Northern Tier*, RAND Report R-2417/1-AF/FF (RAND: Santa Monica, Calif., 1980), pp. 13–18, 30–31.

encompassed strategy, 'operational art' and tactics. Soviet military doctrine, as one US analyst has explained, took the form of the 'official views and recommendations embraced by the *political leadership* regarding the nature of future war, the political and military goals of war, [and] the likely methods with which future war will be fought . . .'.[7] Doctrinal statements were sifted both inside and outside the country for clues to the development of Soviet foreign and defence policy, and they were understood to affect defence industrial and force planning decisions directly.

The determining importance still attached to military doctrine is apparent in the discussions that have taken place in most of these states about the restructuring and management of defence industries. The new doctrines are seen as providing a medium-term planning framework for these industries, albeit one that is strongly influenced by the financial resources available to governments.[8]

As part of the present study investigating changes in the Central and East European arms industries, this chapter looks at the doctrinal determinants of the demand side of defence industrial production. It describes the general features of post-Warsaw Pact military doctrines, focusing on their implications for the size and composition of 're-nationalized' armed forces. It first examines the national threat assessments underlying emerging doctrines. It then looks at how new doctrines are shaping the development of conventional military capabilities to address these threats. The aim here is not to assess the changing demand for specific mixes of weapons and equipment. Rather, it is to highlight the main trends and directions in the doctrinal transitions under way as they affect future general-purpose force structures and levels.

II. National threat perceptions and conflict scenarios

The new military doctrines taking shape among the former Warsaw Pact member states guide force structure and operational planning decisions in accordance with underlying assessments of military

[7] Meyer, S. M., *Soviet Theatre Nuclear Forces, Part I: Development of Doctrine and Objectives*, Adelphi Paper no. 187 (IISS: London, 1984), pp. 3–4 (emphasis in original). See also 'Introduction', eds H. F. Scott and W. F. Scott, *The Soviet Art of War: Doctrine, Strategy and Tactics* (Westview Press: Boulder, Colo., 1982), pp. 4–9.

[8] Vadim I. Vlasov, Russian Federation Ministry of Defence, at the SIPRI workshop on The Future of the Defence Industries of Central and Eastern Europe, 29–30 Apr. 1993.

threats to national security; they lay out a set of prescriptions speci-
fying how armed forces should be structured and employed to
respond to recognized threats. The relationship between national
threat perceptions and force postures is not a wholly deterministic
one, however, since military planning decisions in all of these states
are affected by the severe downward pressures on defence budgets
and the general dislocations attending the transition from state socialism to market economies.[9]

The development of independent national threat assessments
reflects the liberation of defence planners across Central and Eastern
Europe from the ideological baggage of their Warsaw Pact heritage.
In tandem with the purging of the Communist Party apparatus from
the armed forces, they have abandoned the 'internationalist' tenets of
Soviet-imposed doctrine that formed the ideological basis of
Moscow's domination of Warsaw Pact political and military structures.[10] In all these countries, national doctrines have been shorn of
the underlying ideological characterizations of malign Western intentions that drove Soviet threat perceptions and planning for war with
NATO.[11] Indeed, most of the former Warsaw Pact states are now
pushing for membership in the erstwhile 'inimical alliance'.

With the breakdown of the relative stability of European politico-
military arrangements fostered by the old bipolar division of the con-
tinent, national leaders confront an array of new potential military
threats and worrying conflict scenarios. There is general agreement
among them that the proliferation of domestic instabilities and the
resurfacing of latent national antagonisms across the old socialist
military bloc have expanded the range of plausible war contingencies.
Considerable differences emerge, however, when it comes to their
assessments of the nature and likelihood of future conflicts and their
prescriptions for the military means needed to address them.

[9] For a discussion of the relationship between threat perceptions and emerging military
structures in the former Soviet republics, see Allison, R., *Military Forces in the Soviet
Successor States*, Adelphi Paper no. 280 (Brassey's: London, 1993), esp. pp. 72–74.

[10] Jones, C., 'National armies and national sovereignty', eds D. Holloway and J. Sharp,
The Warsaw Pact: Alliance in Transition? (Macmillan: London, 1984), pp. 96–97.

[11] Lange, P. H., 'Understanding military doctrine', ed. L. Valki, *Changing Threat
Perceptions and Military Doctrines* (Macmillan: London, 1992), pp. 14–15.

The Visegrad Group

The post-communist leaderships of the Visegrad Group (the Czech Republic, Hungary, Poland and Slovakia) share broadly similar understandings of their national security interests and the nature of threats to them. They are concerned primarily with new threats arising from the concatenation of internal troubles in the region rather than with direct external military challenges. The threat assessments underlying their new doctrines are expressed in two principal conflict scenarios.

One conflict scenario about which military spokesmen in these countries articulate considerable concern is that of war arising from disputes within or between neighbouring states.[12] The accumulated legacies of unsettled border and territorial grievances, allegations of mistreatment of ethnic kin, and persistent trade and other economic friction are seen as creating a volatile brew of instabilities in the region that can unexpectedly spill over into armed conflicts of unforeseeable intensity and duration. Such conflicts could remain local but could also escalate into large-scale conventional war, particularly if extra-regional states became embroiled in the fighting.[13]

A related conflict scenario derives from concerns that the dissolution of the WTO has created a 'security vacuum' in Central Europe. In particular, officials have expressed fears about a possible resurgence of Russian neo-imperialist aspirations and Moscow's resort to coercion to re-establish a sphere of special influence in the region. Sensitive to the unsettled political situation in Russia, governments of Visegrad Group countries have been for the most part circumspect in addressing this scenario.[14] The chronic instability to the east does, however, cast a long shadow over regional security deliberations. It is clearly a prime mover behind the Visegrad states' decisions to participate in NATO's Partnership for Peace programme

---

[12] de Weydenthal, J., 'Poland's security policy', Radio Free Europe/Radio Liberty, *RFE/RL Research Report*, vol. 2, no. 14 (2 Apr. 1993), p. 31.

[13] Statement by Gen. János Deák, Chief of the General Staff of the Armed Forces of the Republic of Hungary, at the Second Seminar on Military Doctrine, Vienna, 9 Oct. 1991; Koziej, S., 'Poland's defence policy', *Journal of Slavic Military Studies*, vol. 6, no. 1 (June 1993), pp. 152–53.

[14] Szayna, T., *The Military in a Postcommunist Czechoslovakia*, RAND Note N-3412-USDP (RAND: Santa Monica, Calif., 1992), pp. 88–89.

and to push for eventual full membership in NATO and other Western security bodies.[15]

Consistent with foreign policies aimed at promoting greater co-operation and stability, the new military doctrines of these states do not identify specific countries or coalitions against which national armed forces are directed. Poland, for example, has articulated a 'no *a priori* enemy' doctrine that lacks a specific threat image; Hungary has elaborated a 'home defence' doctrine that considers no state to be a 'potential enemy or adversary'.[16]

These doctrines are based on an 'elliptical' or 'all-round' defence concept, in which planning for the armed forces is not anchored in well-defined threats. Within this framework, conflicts are seen as coming from a variety of directions; accordingly, forces are not to be structured and deployed along specific axes.[17]

Russia

The November 1993 military doctrine similarly declines to identify specific adversaries that Russian troops might face. Outlining the contents of the new doctrine at a press conference in Moscow, Defence Minister Pavel Grachev emphasized that 'the Russian Federation does not treat any state as its enemy'.[18]

Grachev stated that Russia pledges never to use military force against any state, except in self-defence and—adding a significant qualification—'in conjunction with the guarantees not to use nuclear weapons against signatory states to the Nuclear Non-proliferation Treaty (NPT) of 1 July 1968 which do not possess nuclear weapons'. The latter statement marks a departure from the Soviet policy on the non-first use of nuclear weapons, at least at the declaratory level. It is widely seen as a warning directed against Ukraine, which has yet to

---

[15] For a presentation of security policy views from the Visegrad states, see the report published by the Windsor Group, *NATO: The Case for Enlargement* (Institute for Defence and Strategic Studies: London, Dec. 1993). See also Reisch, A., 'Central Europe's disappointments and hopes', *RFE/RL Research Report,* vol. 3, no. 12 (25 Mar. 1994), pp. 18–37.

[16] Lachowski (note 5), p. 502.

[17] Clarke, D., 'A realignment of military forces in Central Europe', *Report on Eastern Europe,* 8 Mar. 1991, pp. 41–43.

[18] The full text of the military doctrine approved by the Russian Security Council will not be published, although some parts of it have appeared in the press. For Grachev's press conference remarks announcing the new doctrine, see 'Grachev outlines new military doctrine', Radio Moscow, 3 Nov. 1993, in Foreign Broadcast Information Service, *Daily Report–Central Eurasia (FBIS-SOV),* FBIS-SOV-93-212, 4 Nov. 1993, pp. 34–36.

ratify the NPT and which has claimed ownership over the former Soviet strategic nuclear weapons based on its territory.[19] It may also reflect a judgement in Moscow that a militarily weakened Russia must now rely more on nuclear weapons to deter a potential adversary such as China.[20]

The new doctrine unveiled by Grachev does still pay attention to traditional scenarios of large-scale conflicts, although it is predicated on the notion that the possibility of 'world nuclear or conventional war being unleashed . . . has been considerably diminished'.[21] According to an earlier draft of the doctrine, such conflicts may arise either with the escalation of local wars—whether aimed against Russia or breaking out in regions adjacent to Russia's borders—or after a 'prolonged threat period' that involves general mobilization.[22]

It seems clear, however, from the Defence Minister's presentation of the doctrine and the discussions surrounding it that senior defence planners are most concerned about the proliferation of local wars or low-intensity conflicts along the periphery of Russia. Grachev has asserted that 'Local wars are the main threat to peace, and the possibility of their eruption in certain regions is growing'.[23]

Russian defence officials now identify the main near-term dangers for the country's security arising in 'the south'.[24] A primary threat is seen as the spread of Iranian- and Afghan-propagated Islamic fundamentalism among Russia's estimated 20 million Muslims. Moscow's anxieties are heightened by the related fear that ethnic and national conflicts simmering in the Caucasus will spill over into areas of southern Russia, where worrying disintegrative trends are already evident. According to one US analyst of Russian military affairs, 'the south of Russia is thus seen as a firebreak, with a strong presence

[19] See note 18.
[20] The change may also have been intended to discourage other former Soviet republics and former WTO member states from joining NATO. Schmemann, S., 'Russia drops "first-use" vow on atom arms', *International Herald Tribune*, 4 Nov. 1993, pp. 1 and 6; Lockwood, D., 'Russia revises nuclear policy, ends Soviet "no-first-use" pledge', *Arms Control Today*, vol. 23, no. 10 (Dec. 1993), p. 19.
[21] See note 18.
[22] Fitzgerald, M., 'Russia's new military doctrine', *Naval War College Review*, vol. 46, no. 2 (spring 1993), p. 35.
[23] Quoted in Sneider, D., 'New Russian military doctrine raises Western suspicions', *Defense News*, 8–14 Nov. 1993, p. 34.
[24] Allison (note 9), p. 22.

being required there in order to stop threats from Asia along the border'.[25]

The 1993 doctrine assigns to the Russian armed forces a number of new missions not found in previous doctrines. An important—and highly politicized—mission codified in the new doctrine is the use of the army to quell disturbances and conflicts *within* the borders of the Russian Federation.[26] There is also considerable attention given to Russia's participation in international peacekeeping operations.[27]

In addition, the new doctrine includes the provision that under certain circumstances Russian armed forces might initiate military action in adjacent countries in order to protect the rights and interests of ethnic Russians residing outside the borders of the Russian Federation. Grachev and other senior MOD officials have identified the protection of ethnic Russians abroad as a bona fide mission of the Russian military.[28] This scenario coincides with the unilateralist tendencies evident in Russia's approach to its self-appointed special role as peacekeeper across the territory of the former Soviet Union. It is one that has aroused considerable disquiet in the West and among Russia's neighbours, since it seems to justify Russian military intervention in the other former Soviet republics comprising the so-called 'near abroad'.[29]

Ukraine

The contentious parliamentary debates leading up to the Rada's approval on 19 October 1993 of a national military doctrine revealed that the majority of deputies see a resurgence of Russian expansionism—in the form of either Moscow-supported separatist movements or direct military attack—as posing the gravest danger to Ukrainian sovereignty and independence. Although the new doctrine does not

[25] Lepingwell, J., 'Restructuring Russia's military doctrine', *RFE/RL Research Report*, vol. 2, no. 25 (18 June 1993), p. 18. See also Solovyev, V., 'Pavel Grachev builds up Russia's defense in a new fashion, sees threat where it formerly did not exist', *Nezavisimaya Gazeta*, 7 May 1993, p. 1, in FBIS-SOV-93-088, 10 May 1993, pp. 38–39.

[26] For a discussion of Russian military views on internal operations, see Taylor (note 3).

[27] See note 18.

[28] Gordon, M., 'As its world view narrows, Russia seeks a new mission', *International Herald Tribune*, 29 Nov. 1993, pp. 1 and 7.

[29] See, for example, the interview with Ivan Plyushch, Chairman of the Supreme Council of Ukraine, Kiev UNIAR (in Ukrainian), 5 Nov. 1993, in 'Plyushch, Kravchuk on Russian military doctrine', FBIS-SOV-93-214, 8 Nov. 1993, p. 58; see also Odom, W., 'A new Russian empire may be coming', *International Herald Tribune*, 26 Oct. 1993, p. 7.

specifically identify Russia as a likely foe, its final version includes an amendment stating that 'Ukraine will consider its potential adversary to be any state whose consistent policy constitutes a military danger to Ukraine'.[30]

Senior defence ministry officials in Kiev remain divided over the priority to be given to contingency planning for large-scale conflict with Russia.[31] A hierarchy of threats does seem to be emerging in Ukrainian defence planning discussions, however, in which relatively less attention is paid to scenarios of local wars arising from disputes with neighbouring states (e.g., Moldova or Romania). Against the background of insistent demands from nationalist deputies in parliament for a re-orientation of the armed forces, the new military doctrine reflects Kiev's gradual move away from a broadly neutral *tous azimuts* force posture towards one structured and deployed to meet potential threats emanating from Russia.[32] At least in the medium term, however, plans do not call for significant redeployments of forces along operational axes in northern and eastern Ukraine.

III. A survey of selected military doctrines

The Visegrad Group

The new military doctrines that have been elaborated in the Czech Republic, Hungary, Poland and Slovakia feature similar sets of core assumptions about the kind of armed forces needed in the transformed European security environment. They are predicated on the notion that military capabilities should be adequate to defend the national territory while at the same time not being so great as to threaten the security of other states; the defensive character of the armed forces should be evident in their size, structure and deployment patterns. The importance of transparency, openness and predictability in military

---

[30] Kiev UNIAR (in Ukrainian), 20 Oct. 1993, in FBIS-SOV-93-201, 20 Oct. 1993, p. 71. Many nationalist deputies had objected to an earlier draft of the doctrine which stated that 'Ukraine does not consider any state to be its adversary'. Markus, U., 'Recent defense developments in Ukraine', *RFE/RL Research Report*, vol. 3, no. 4 (28 Jan. 1994), p. 29.

[31] Kulida, S., 'Ukraine's likely enemy—Russia', *Shlyakh Peremohy* (Lvov), 5 Mar. 1994, p. 5, in FBIS-SOV-94-046, 9 Mar. 1994, pp. 31–32; Vasyanovych, P., 'On military and economic security', *Kyyivska Pravda*, 25 Nov. 1993, in FBIS-SOV-93-230, 2 Dec. 1993, p. 52.

[32] Markus (note 30).

planning and activities is another common thread running through these doctrines.

Defence sufficiency and force postures

One of the key tenets of force planning is the notion that national armed forces should be trained, equipped and deployed according to the principle of sufficiency. In defence ministries across Central Europe, sufficiency is essentially defined in terms of 'defensive defence'. It connotes a force posture that is capable of repelling an attack, but that is incapable of conducting massive offensive operations against the territory of an aggressor.[33] It is based upon a military strategy of offensive self-denial: national armed forces are to be configured and deployed to halt an attack, hold against further penetration of national territory and then push the aggressor back to the border. However, these forces will not be able to conduct large-scale operations beyond the confines of the national territory.

The restructuring programmes under way across Central Europe are accordingly aimed at creating forces with only tactical counter-offensive capabilities. Officials in these countries are keen to stress that concrete changes are being made to bring force postures fully into line with declared defensive principles. In Poland, for example, the fire-power potential of armoured and mechanized infantry formations, the maximization of which was a key element in Warsaw Pact force planning, is being reduced in favour of greater unit manœuvrability for defensive operations. (A fire-power advantage is considered necessary to offset the higher expected rate of attrition involved in conducting offensive operations.) The defensive reorientation of military doctrines has also percolated down to the tactical level; field training exercises emphasize all-round defence of important garrison regions, battles on successive defensive-delaying lines, harassing actions by rapid reaction forces, and so on.[34]

The definition of sufficiency in terms of defensive defence would seem to provide a natural standard for the size and peacetime activi-

---

[33] One Polish official has defined sufficiency as 'an ability to defend our territory and a real inability to wage aggressive operations'. Statement by Janusz Onyszkiewicz, Deputy Minister for Defence of the Republic of Poland, at the Second Seminar on Military Doctrine, Vienna, 8 Oct. 1991.

[34] Koziej, S., 'Main problems of operational art and tactics of Poland's ground forces in the 1990s', *Journal of Slavic Military Studies*, vol. 5, no. 4 (Dec. 1992), pp. 570–71.

ties of national armed forces. It implies the abandonment of concern with maintaining numerical parities and matching the military capabilities of a putative adversary; decisions about the procurement and deployment of forces are to be determined by the requirements to defend national territory.

From a force planning perspective, however, the above requirements are not easy to spell out in concrete terms. The lack of explicit threat-based force planning makes it difficult to identify and support planning objectives for future core military capabilities. It provides planners with little practical guidance in deciding the appropriate size and composition of the armed forces, rate of modernization, degree of combat readiness to be maintained, and so on. Such decisions cannot be reduced to a technical formula that is immune to analytic dispute.

Nor are the decisions immune to political suspicion about underlying intentions. Given that defensive and offensive force postures are not discrete categories but fall along a continuous spectrum, the reconfiguration of armed forces for an era of sufficient defence is still laden with ambiguities that can give rise to security dilemma anxieties (that is, the propensity for the defensive preparations of one state to appear threatening to its neighbours). Hungary's decision to modernize its air defence force with the acquisition of Russian-manufactured MiG-29 fighter-bombers, for example, prompted Slovakia to announce that it intended to acquire five MiG-29s from Russia.[35]

Russia

The professional military remains the dominant actor in planning and implementing changes in the basic structure and organization of Russia's conventional forces. Not surprisingly, then, Russian military thinking shows many continuities with its Soviet antecedents; some of the favourite themes of the old Soviet General Staff survive in Russia's new military doctrine. The balance of opinion among senior defence officials over the acceptability of far-reaching doctrinal reforms is fluid, however, and the armed forces are clearly being restructured in light of 'new realities'.

In response to proposals put forward by a cadre of civilian analysts to build national armed forces possessing only limited counter-

[35] Robinson, A., 'Slovak PM personifies republic's image problems', *Financial Times*, 12 Aug. 1993, p. 4.

offensive capabilities, Russian military planners have consistently stressed the need to maintain flexibility for the employment of forces.[36] The new military doctrine calls on the armed forces to be able 'to conduct both defensive and offensive operations, in conditions of the massive use of present and future weapons'.[37] In this regard, the doctrine continues the steady retreat led by senior officers away from a Gorbachev-era principle of reasonable sufficiency embracing a defence-dominant theatre strategy and force posture.[38]

The November 1993 military doctrine continues to mandate a substantial defence effort based on a mass mobilization capacity, although it lacks the previous draft's imputation that Russia will continue to use NATO and the United States as a strict yardstick for measuring the quantitative elements of its military efforts.[39] Some Russian analysts have argued that, given the country's sprawling expanse and pivotal geostrategic position in Eurasia, prudent military planning mandates preserving relatively large standing forces and cadres of mobilization reserves; such forces are needed not only as hedge against an unlikely worst-case conventional war contingency but also in case simultaneous contingencies should arise (that is, if Russia were to find itself engaged in fighting in more than one theatre at the same time).[40]

A top priority for the reorganization of the armed forces is the formation of centrally based air-mobile units that can be rapidly deployed to any threatened axis or to neighbouring trouble spots.[41] Since Moscow lacks the manpower and other resources to mount a deep perimeter defence along the entire border, mobile forces are envisaged as complementing permanent readiness covering units stationed in-

[36] See, for example, Gareev, M. A., 'On military doctrine and military reform in Russia', *Journal of Slavic Military Studies*, vol. 5, no. 4 (Dec. 1992), p. 549.

[37] See note 18.

[38] For additional background to the internal Russian debate on sufficiency, see Meyer, S. M., 'The sources and prospects of Gorbachev's new political thinking on security', *International Security*, vol. 13, no. 2 (autumn 1988), pp. 124–63; see also Silverman, W., 'Talking "sufficiency" in the Hofburg Palace: the Second Seminar on Military Doctrine', *Arms Control Today* , vol. 21, no. 10 (Dec. 1991), p. 14.

[39] Dick, C. J., 'Initial thoughts on Russia's draft military doctrine', *Journal of Slavic Military Studies*, vol. 5, no. 4 (Dec. 1992), pp. 552–66.

[40] Arbatov, A., 'Russian foreign policy priorities for the 1990s', eds S. E. Miller and T. Pelton Johnson, *Russian Security after the Cold War: Seven Views from Moscow*, CSIA Studies in International Security Affairs No. 3 (Brassey's: Washington DC, 1994), p. 37.

[41] See Vladykin, O., 'Russia's mobile forces: they are being created in response to the dictates of the times', *Krasnaya Zvezda*, 18 Dec. 1992, p. 2, in FBIS-SOV-245, 21 Dec. 1992, pp. 42–44.

theatre. Their mission is to localize incipient conflicts and prevent their escalation. In addition, during wartime or periods of threat, these forces (along with other standing units) would support the mobilization and deployment of the country's strategic reserves.[42]

From a force planning perspective, the new doctrine appears to go some way towards resolving an ambiguity in the draft doctrine published in May 1992. The earlier draft specified that the Russian armed forces should be prepared to meet both a large-war contingency, presumably involving NATO, and a smaller 'ethnic conflict contingency'; it did not indicate, however, which of the two contingencies—which require substantially different military capabilities—should be given priority in the reorganization of the armed forces.[43] The new doctrine's emphasis on fighting low-intensity conflicts in or near Russia's borders suggests that greater importance is now attached to the formation of lighter mobile forces rather than to traditional Soviet tank-heavy combined arms formations.[44]

Although the mobile forces will form a core element of the armed forces, they will by no means constitute the entire Russian force posture. Senior Defence Ministry officials have stressed the need to maintain forces able to meet a range of contingencies, including those involving large-scale combat operations.[45] Considerable emphasis is being given to the creation of more flexible and manœuvrable forces equipped with latest-generation high-technology weapon systems comparable to those in Western arsenals. The ground forces are to be reduced and radically restructured, while the development of the air force emerges as a key area.[46]

In addition, the Defence Ministry has assigned priority to the acquisition of long-range, precision-guided conventional munitions, electronic warfare assets and advanced reconnaissance and command,

[42] Allison (note 9), pp. 23–24. Falichev, O., 'Continuation of the military reforms is a priority state task', *Krasnaya Zvezda*, 10 Mar. 1994, p. 1, in FBIS-SOV-94-049, 14 Mar. 1994, p. 28.

[43] Lepingwell (note 25), p. 18. See also Blank, S., 'New strategists who demand the old economy', *Orbis*, vol. 36, no. 3 (summer 1992), pp. 365–78.

[44] According to Deputy Defence Minister Andrey Kokoshin, in current conditions the latter formations, which formed the backbone of the Soviet army, are 'dinosaurs from World War II'. Quoted in Agapova, Y., 'Before you form an army you should know what it is for, expert Andrei Kokoshin believes', *Krasnaya Zvezda*, 17 Mar. 1992, pp. 1–2, in FBIS-SOV-92-053, 18 Mar. 1992, p. 27.

[45] See, for example, Grachev, P., 'Drafting a new military doctrine', *Military Technology*, no. 2 (1993), pp. 14–15.

[46] Allison (note 9), p. 23.

control, communications and intelligence (C³I) systems.[47] The performance of the USA's so-called 'reconnaissance-strike complex' in the 1991 Persian Gulf War is seen by some Russian observers as confirming the potential decisiveness of these systems, especially in the critical stage at the onset of a conflict.[48]

IV. Changes in force structures and levels

Central Europe

The armed forces of the former NSWP member states in Central Europe are in the midst of comprehensive transformation aimed at yielding smaller, more modern defensive forces. The changes, some of which were set in train during the waning years of communist rule, involve reductions and restructuring of the core military capabilities to bring them into line with defensively oriented military doctrines based upon the principle of defence sufficiency.

Although these changes are proceeding according to independently formulated national programmes, there are a number of common strands running through them. One of these is the ambition of national defence ministries to reduce or eliminate their most obvious residual offensive capabilities. Poland, for example, has reduced the number of its pontoon bridge assault units and replaced large mobile logistics formations with stationary bases as part of its restructuring programme entitled 'Army of the 1990s'.[49] Hungary announced in November 1990 that it was scrapping its inventory of Scud-B and FROG-7 tactical ballistic missiles.[50]

The restructuring of residual offensive force postures also involves more comprehensive measures. The Czech Republic, Hungary, Poland and Slovakia have all undertaken to reorganize their Warsaw Pact-era motorized rifle and armoured divisions, which were pat-

[47] Grachev (note 45).

[48] See Kaufman, S., 'Lessons from the 1991 Gulf War and Russian military doctrine', *Journal of Slavic Military Studies*, vol. 6, no. 3 (Sep. 1993), pp. 377–86; see also Fitzgerald (note 22), pp. 36–37.

[49] Ripley, T., 'The Polish armed forces in the 1990s', *Defense Analysis*, vol. 8, no. 1 (1992), pp. 88–90. See also Presentation by Maj.-Gen. Zdzislaw Stelmaszuk, Chief of the General Staff of the Polish Armed Forces, at the Second Seminar on Military Doctrine, Vienna, 11 Oct. 1991.

[50] Reisch, A., 'The Hungarian Army in transition', *RFE/RL Research Report*, vol. 2, no. 10 (5 Mar. 1993), p. 42.

terned on the tank-heavy Soviet model, into lighter mechanized infantry and armoured brigades. These units are maintained at lower overall readiness levels than their Warsaw Pact predecessors.[51]

In addition, consistent with doctrines emphasizing the 'all-around' defence of national territory, planners in these states have now largely completed an extensive and costly eastward redeployment of their armed forces. The bulk of these forces had been deployed in the western parts of Poland, Czechoslovakia and the German Democratic Republic as part of the Warsaw Pact's emphasis on deep echelons of combat formations geared toward a rapid, Soviet-led offensive against NATO.[52] These redeployments were also spurred by the apprehension of governments in the region that the chronic instability on the territory of the former Soviet Union could unleash a massive westward flow of refugees.[53]

Military planners across Central and Eastern Europe have assigned priority to improving the ability of ground forces to respond quickly to local wars and low-intensity conflicts; these changes have been particularly urgent in the countries adjacent to the former Yugoslavia, where fighting has occasionally threatened to spill over the borders. Defence ministries have assigned priority to creating a core of centrally-based air-mobile rapid reactions units at the battalion and brigade level.[54]

The emphasis on mobility and rapid reaction will require an upgrading of ground forces' reconnaissance and anti-tank capabilities. It has been estimated that the Hungarian Army, for example, needs an additional 30–40 per cent more anti-tank weapons.[55]

In addition, defence ministry plans in the Czech Republic, Hungary, Poland and Slovakia give high priority to the enhancement of national air defence capabilities. The collapse of the extensive Warsaw Pact air defence network, which was integrated under the commander of the Soviet air defence force (PVO Strany), has left these countries with a weakened ability to defend their air space. Plans call for the

[51] Clarke (note 17), p. 43.

[52] As an example of this deployment imbalance, 8 Warsaw Pact divisions were based in parts of the Czech Republic, whereas only 2 low-readiness divisions were based in Slovakia. Clarke (note 17), pp. 41–43.

[53] Szayna (note 14), pp. 62–63.

[54] Western European Union, *Defence: Central Europe in Evolution*, WEU document 1336, 5 Nov. 1992, pp. 175–77.

[55] Reisch (note 50), p. 46.

acquisition of new aircraft, as well as additional anti-aircraft missile batteries and upgraded air defence radars.[56]

Senior officials have indicated that they would like to purchase Western-made equipment compatible with NATO standards, although this is not feasible due to downward pressures on procurement budgets. It is clear, however, that considerable changes will have to be made in the organization and equipment of all the armed forces in the region if they are to meet NATO's requirements.[57]

Ukraine

Ukraine became the first of the former Soviet republics to create independent armed forces when a presidential decree issued in January 1992 placed all non-strategic troops and equipment holdings of the Soviet Carpathian, Odessa and Kiev Military Districts under the jurisdiction of the country's new Defence Ministry.[58] These forces, which under Soviet theatre war plans constituted the first strategic echelon directed against NATO, were considered to be among the most modern and combat-capable units in the Soviet armed forces. Ukrainian defence ministry officials have complained, however, about the state of their inherited equipment holdings. [59]

Despite the absence of a consensus on an overarching operational concept, a comprehensive restructuring of the Ukrainian armed forces is well under way. The aim is to create a smaller, more modern army with greater operational mobility and rapid-reaction capabilities. The old Soviet army group formations are being replaced by a corps/ brigade structure. Although the CFE Treaty permits Ukraine to maintain extensive treaty-limited equipment (TLE) holdings (see table 2.1), much of this hardware is likely to be placed in storage as the force structure is pared down.[60]

[56] See the interview with Hungarian Air Force Inspector General János Urbán in *Jane's Defence Weekly*, vol. 18, no. 23 (5 Dec. 1992), p. 56.

[57] de Weydenthal, J., 'Poland builds security links with the West', *RFE/RL Research Report*, vol. 3, no. 14 (8 Apr. 1994), p. 30.

[58] Sauerwein, B., 'Rich in arms, poor in tradition: the Ukrainian armed forces', *International Defense Review*, no. 4 (1993), p. 317; Zaloga, S., 'Armed forces in Ukraine', *Jane's Intelligence Review*, Mar. 1992, p. 133.

[59] One official has estimated that 80% of Ukraine's rocket artillery, 50% of its air force equipment and 100% of its anti-tank weapons are obsolete. Sauerwein (note 58), p. 318.

[60] Sauerwein (note 58), p. 318; Allison (note 9), pp. 42–43.

Table 2.1. Former Warsaw Pact CFE Treaty ceilings[a]

CFE party	Tanks	ACVs	Artillery	Aircraft	Helicopters	Total
Bulgaria	1 475	2 000	1 750	234	67	**5 526**
Czech Rep.	957	1 367	767	230	50	**3 371**
Hungary	835	1 700	840	180	108	**3 663**
Poland	1 730	2 150	1 610	460	130	**6 080**
Romania	1 375	2 100	1 475	430	120	**5 500**
Slovakia	478	683	383	115	25	**1 684**
Russia[b]	6 400	11 480	6 415	3 450	890	**28 635**
Ukraine[b]	4 080	5 050	4 040	1 090	330	**14 590**

[a] In the Atlantic-to-the Urals (ATTU) zone of application

[b] Represents the CFE national ceilings agreed to by the former Soviet republics in the Joint Declaration signed at the CIS summit meeting in Tashkent, Uzbekistan, on 15 May 1992. (The three Baltic states—Estonia, Latvia and Lithuania—did not participate.)

Source: Sharp, J. M. O., 'Conventional arms control in Europe', SIPRI, *SIPRI Yearbook 1993: World Armaments and Disarmament* (Oxford University Press: Oxford, 1993), p. 609.

The process of building national armed forces suited to the country's defence needs has been hampered by drastic reductions in military spending. Low pay and lack of housing have prompted an exodus of experienced officers and skilled personnel from the armed services.[61] Combat readiness has been further undermined by training shortfalls[62] and shortages of fuel and spare parts.[63] In addition, the parlous state of the defence budget has made the Defence Ministry's ambitious plans for a highly modern force based on advanced technology weapon systems clearly unrealistic.[64]

Military planners in Kiev have also been forced to scale back ambitious initial force level plans significantly as the heavy costs

[61] Radetskyy, V., 'Where did you see an army without a budget?', *Ukrayinska Hazeta*, no. 2, 30 Jan.–2 Feb. 1994, p. 4, in FBIS-SOV-94-015, 24 Jan. 1994, p. 75. Personnel problems have been aggravated by widespread draft evasion and a bitter dispute about Russian and Ukrainian service oaths.

[62] According to one report, the Ukrainian Army did not hold a single exercise at the regimental level during 1993. Vorotynskyy, I., 'Minister Radetskyy at the crossroads', *Post-Postup* (Lvov), 4 Mar. 1994, p. 2, in FBIS-SOV-94-052, 17 Mar. 1994, p. 24.

[63] The disruption of integrated defence production links with Russia has further eroded readiness levels and force modernization. Radetskyy, V., 'Professionalism, discipline and order', *Armiya Ukrayiny*, 26 Jan. 1994, pp. 1–2, in FBIS-SOV-94-022, 2 Feb. 1994, p. 27; Sauerwein (note 58), p. 318.

[64] Radetskyy (note 61).

involved in restructuring and maintaining such a force have begun to be better appreciated.[65] According to Defence Minister Vitaliy Radetskyy, at the beginning of 1994 the armed forces numbered approximately 533 000 servicemen. Plans call for reductions in total personnel strength to 300 000–350 000 troops by the year 2000.[66] This is well below Ukraine's CFE-1A entitlement of 450 000 (see table 2.1 above). One Western military analyst has argued that even this reduced force size will exceed feasible military outlays and that it could drop to as low as 100 000–150 000 men.[67]

Russia

Russian military planners are confronted with a number of serious practical problems in shaping the post-Soviet force structure. A key challenge is to restore military cohesion and to overcome the loss of important elements of the former Soviet armed forces. The splintering of the USSR has particularly disrupted integrated early warning, air defence and logistical support systems.[68] The force structure Russia inherited consists of a relatively high percentage of less combat-ready units; a large proportion of high-readiness units equipped with the most advanced weapons remains outside the Russian Federation, although this situation is changing somewhat with the return of Russian units from Germany and other parts of Europe.

Russian military planners have inherited a residual force structure oriented to operations for defending the periphery of the territory of the former Soviet Union. Some Russian forces are stationed on the territory of what are now independent states. They are also badly deployed with respect to the country's new geopolitical position and new regions of conflict.[69] Few combat formations are stationed, for example, along the volatile southern rim of the Russian Federation.

Current MOD plans call for far-reaching changes in the command structure of the Russian armed forces. Of fundamental importance, the basis of the Soviet mobilization and military command structure—the

---

[65] See the interview with Col.-Gen. Konstantin Morozov, Ukrainian Minister of Defence, in *Jane's Defence Weekly*, 14 Aug. 1993, p. 32.

[66] Radetskyy, (note 61).

[67] Allison (note 9), p. 42.

[68] Allison (note 9), p. 28; see also Grachev (note 45).

[69] Donnelly, C., 'Evolutionary problems in the former Soviet armed forces', *Survival*, vol. 34, no. 3 (autumn 1992), p. 34.

military district—is being discarded, since many of these districts now fall outside of the territory of the Russian Federation or are no longer relevant to the country's post-Soviet security situation. The intention is to replace the old district system with a system of geographically designated regional commands by the year 2000. The process of forming such regional groupings, which are likely to combine combat units, logistics troops, military transport aviation and airmobile forces, is already under way in the Russian Far East and the North Caucasus, the latter region having been assigned highest priority for deployments of new personnel and equipment.[70]

In addition, consistent with its new emphasis on mobile forces designed for fighting smaller conflicts in or near Russia's borders, the MOD has announced plans to reorganize the present force structure by creating a Mobile Forces Command that will eventually consist of approximately 100 000 troops. This command comprise two components: a mobile force and a rapid deployment force.[71] The remaining ground forces are to be converted from the present Soviet-era division/army structure into a predominantly brigade/corps structure in order to increase the number of combat-ready units.[72]

The re-allocation of the Soviet allotment of CFE TLE agreed between the successor states at Tashkent in May 1992 leaves Russia with extensive equipment holdings, although much of this hardware is obsolete. The Russian Government has officially raised the issue of revising certain geographic sub-limits contained in the CFE Treaty to enable the relocation of forces to the southern regions of the Russian Federation (for example, the North Caucasus).[73]

Under the terms of the CFE-1A Agreement on personnel strength (signed by Russia in July 1992), the ceiling on manpower deployed by the Russian armed forces (excluding the naval forces and the strategic rocket forces) west of the Ural mountains is 1 450 000 (see table 2.2). In accordance with the 1992 Basic Law on Defence, which stated that no more than 1 per cent of the population may be serving in the armed forces by 1 January 1995, the total size of the Russian armed forces would be reduced to no more than 1 500 000 troops by that time.

[70] Allison (note 9), pp. 29–30.
[71] Woff, R., 'Russian mobile forces 1993–95', *Jane's Intelligence Review,* Mar. 1993, pp. 118–19; Lepingwell (note 25).
[72] Allison (note 9), p. 29.
[73] Lepingwell (note 25).

Table 2.2. CFE-1A manpower limitations

State	Ceilings	Holdings
Bulgaria	104 000	99 404
Czech Republic	93 333	110 010
Hungary	100 000	76 226
Poland	234 000	273 050
Romania	230 000	244 807
Slovakia	46 667	55 005
Russia	1 450 000	1 298 299
Ukraine	450 000	509 531

Source: Concluding Act of the Negotiation on Personnel Strength of Conventional Armed Forces in Europe (CFE-1A Agreement), 10 July 1992, Helsinki.

In December 1993, Grachev indicated that the MOD would not abide by the ceiling imposed by the previous Russian parliament, which he claimed was too low to meet the country's defence needs; the total manpower level of the Russian armed forces is to be maintained at approximately 2 million.[74] It seems doubtful, however, whether Russian personnel strength can be held at this level in the light of severe financial constraints and recruiting shortfalls.[75]

V. Concluding remarks

The governments of the former Warsaw Pact member states have embarked upon ambitious programmes of military reform and restructuring to create armed forces that correspond to their national security needs in the transformed post-cold war security environment. All these countries are in the midst of wrenching economic transitions, however, which seriously constrain the financial resources available for restructuring and modernizing residual Warsaw Pact defence establishments. In an era of rapidly dwindling military expenditures, the issue of affordability looms large for the governments of these states. Ultimately, they must match their defence strategies to new realities and to available resources.

[74] ITAR-TASS (in Russian), 29 Dec. 1993, in 'Defense Minister Grachev holds news conference', FBIS-SOV-93-249, 30 Dec. 1993, pp. 37–38.
[75] According to one report, in 1993 the Russian armed services inducted fewer than 22% of the total number of men eligible on the military draft register. Falichev, O., 'The fall '93 draft was very difficult', *Krasnaya Zvezda*, 25 Jan. 1994, p. 1, in FBIS-SOV-94-017, 26 Jan. 1994, p. 18.

3. Military expenditure in transition

Evamaria Loose-Weintraub and Ian Anthony

I. Introduction

In the wake of the disintegration of the cold war system of alliances in Europe, military expenditure trends in the states of Central and Eastern Europe will not emerge until the purpose and functions of military forces in the new conditions are clearly defined[1] and until a process for deciding the level and allocation of public expenditure emerges. Economic constraints have led to reductions in budget allocations for the armed forces across Europe. However, while the trend in military expenditure has been downward in most European countries, the reductions in Western Europe appear gradual and limited compared with those of Central and Eastern Europe.

Allocating resources to rebuilding national armed forces has been subordinated to developing market economies and bringing about a material improvement in living standards. The gap between the expectations created by systemic societal change and economic realities has been identified as the primary security threat in Central and Eastern Europe. Rightly or wrongly, the danger that disaffected populations will be the breeding ground for populist political movements is seen as greater than the danger from the direct application of military power.[2]

The sudden winding up of the Council for Mutual Economic Assistance (CMEA) plunged Central and East European countries—whose financial and trade systems were closely integrated—into an economic crisis. For the past few years, levels of central government expenditure have been decided against the backdrop of high unemployment, falling industrial production and steeply rising inflation.

---

[1] For a discussion of evolving military doctrines in Central and Eastern Europe, see chapter 2.
[2] Maciej Perczynski, Polish Institute of International Affairs, at the SIPRI workshop on The Future of the Defence Industries of Central and Eastern Europe, 29–30 Apr. 1993. For a list of workshop participants, see the appendix.

II. Military expenditure data

It is unlikely that any of the methodologies applied by government or non-government analysts during the cold war gave a clear picture of resource allocation in the WTO.[3] In spite of democratization and greater transparency, calculating military expenditure in Central and Eastern Europe remains a difficult exercise.

There is no generally accepted standard definition of military expenditure, and the resources allocated to the military may not be limited to the sums contained in official defence budgets. Within official defence budget figures there are national variations in accountancy procedures. For example, when constructing a defence budget countries make individual decisions about whether to include or exclude allocations for interior ministry troops, coast guard, border security or other para-military forces, pensions of retired service personnel, and so on.[4]

The largest group of countries to publish standardized military expenditure data is NATO—where the continuous focus on burden-sharing forced the Alliance to develop common methodologies for expressing military expenditure. Consequently, the NATO definition of military expenditure has sometimes been taken as a guideline by other institutions which produce comparative data (including SIPRI). The International Institute for Strategic Studies (IISS) often uses national definitions of military expenditure. Other important sources of information include government submissions to international organizations—notably the International Monetary Fund (IMF) and the United Nations Project on Military Expenditure—each of which has its own definition of military expenditure. As a result it is possible for analysts to be confronted with four or five competing numbers, each of which may purport to represent annual military expenditure and none of which is self-evidently more reasonable than the others.

Since 1989 the budget process has been characterized by greater public awareness (especially among the media) and more openness across Central and Eastern Europe. Ministers and other officials are

[3] The methods used for making estimates and the weaknesses inherent in them are described in Eyal, J. and Anthony, I., Royal United Services Institute, *Warsaw Pact Military Expenditure* (Jane's Publishing Group: London, 1988); and Jacobsen, C., SIPRI, *The Soviet Defence Enigma: Estimating Costs and Burden* (Oxford University Press: Oxford, 1987).

[4] The problems associated with establishing and comparing levels of military expenditure are described in Ball, N., *Security and Economy in the Third World* (Princeton University Press: Princeton, N.J., 1988).

available for interviews and offer public commentaries on defence issues outside a managed news environment for the first time. While this has generated new sources of data, comparative analysis depends on the level of disaggregation and the clarity of explanation that accompanies published data. In fact, military expenditure data are often highly aggregated and unexplained.

Dealing with this problem of lack of comparability and disaggregation is one aim of the North Atlantic Cooperation Council (NACC) being pursued through the 1992 NACC Work Plan for Dialogue, Partnership and Co-operation.[5] As a by-product there may be greater openness and more widespread availability of official information on the military expenditure of Central and East European countries. However, neither the meetings nor the reports on defence budgets produced in the framework of the NACC Work Plan are available outside government circles. The same is true for the military expenditure data exchanged annually by members of the CSCE in the context of the Vienna Document 1992.[6]

Apart from these problems—which are common to the general field of military expenditure analysis—there are specific problems which arise in Central and Eastern Europe. The movement to a market economy together with the simultaneous transformation of the body politic have profound implications for the budget process and the management of public expenditure. Although consultations on how to plan and implement a budget are one element of NACC, the process of developing new procedures will inevitably move at an uneven pace.[7] The greatest progress in introducing new procedures has been made by Hungary and Poland while the greatest difficulties have been experienced by Russia and Ukraine.

Relative price distortions have always characterized the components of military spending in the former WTO countries, and these distortions have become more extreme during the transition to market economies. Some elements of the budget—such as the cost of wages for military personnel—are now valued at market prices. Other elements—notably equipment purchased from the domestic arms indus-

[5] An overview of NATO efforts to educate budget planners in Central and Eastern Europe is given in George, D., 'NATO's economic co-operation with NACC partners', *NATO Review*, Aug. 1993, pp. 19–22.

[6] See Lachowski, Z., 'The Vienna confidence- and security-building measures in 1992', SIPRI, *SIPRI Yearbook 1993: World Armaments and Disarmament* (Oxford University Press: Oxford, 1993), p. 622.

[7] See George (note 5).

try—still have an element of administered pricing. In spite of this residual use of some pricing techniques associated with central planning, the systemic transition currently under way in the economies of Central and Eastern Europe may make it impossible to develop a smooth interface between future budget information and the historical time series based on estimation.

Budget plans established for the year ahead are usually quite different from actual amounts spent by the end of the year. While supplementary budget allocation may occur as an emergency measure in other countries, this practice is routine across Central and Eastern Europe. The allocation of additional funds in excess of budgeted amounts also occurs via subsidies or deficit financing. Extrabudgetary accounts and transfers from state-owned banks to defence enterprises are not accounted for in the defence budget.

The defence establishment (meaning uniformed and civilian elements in the military as well as the defence industry) may also generate revenue outside the defence budget—for example, through the sale of equipment to domestic or foreign buyers.

III. Selected country analyses

Russia

Russia contributed overwhelmingly to the aggregate military expenditure of the former Soviet Union. Russia contributed about two-thirds of Soviet military spending between 1989 and 1991. In 1992 Russia bore most of the costs of troop demobilization compared to the other members of the Commonwealth of Independent States (CIS).

There is near consensus concerning the level of Soviet military expenditure in 1990 and 1991. The figures of 71.99 billion roubles (submitted to the United Nations) for 1990 and 96.6 billion roubles for 1991 are widely accepted.

The future level of Russian military expenditure will be one key determinant of Russian military posture.[8] Therefore, it is an issue of great importance to many countries. Nevertheless, there is no official estimate of Russian military expenditure and in present circumstances few analysts even try to estimate aggregate military spending for the years 1992 and 1993.

---

[8] For a discussion of Russian military doctrine, see chapter 2.

Table 3.1. Estimates of aggregate military expenditure by Russia, 1991–93

All in b. roubles in current prices unless stated otherwise in notes.

Source	1991	1992	1993
Vasiliy Vorobev	80.9[a]	65.5[b]	66.4[b]
Vasiliy Barchuk		93.7[b]	
Andrey Nechaev		632.0	1 550.0[c]
International Institute for Strategic Studies (IISS)		715.7	3 115.5
Russian Parliament			8 011.0

[a] Russia's share in the Soviet military budget.
[b] In 1991 prices.
[c] In 1992 prices.

Sources: Radio Free Europe/Radio Liberty, *RFE/RL News Briefs*, vol. 2, no. 8 (8–12 Feb. 1993), pp. 5 and 7; *RFE/RL News Briefs*, vol. 1, no. 40 (9 Oct. 1992), p. 53; *Military Industrial Complex Newsletter*, no. 8 (Aug. 1993), p. 8. IISS, *The Military Balance 1993–1994* (Brassey's: London, 1993).

The only data for Russia published in any currency other than roubles have been offered by IISS, which estimates expenditure of $39.68 billion in 1992 and $29.12 billion in 1993.[9] This conversion is achieved by applying the official commercial exchange-rate to a rouble figure. As indicated in table 3.1, there are competing estimates of Russian military expenditure in roubles for both 1992 and 1993.

The individuals listed in table 3.1 released aggregate military expenditure figures at times when they were in positions of authority. Lt-General Vasiliy Vorobev was the head of the Defence Directorate within the Ministry of Finance, Vasiliy Barchuk was the Minister of Finance, and Andrey Nechaev was the Minister of Economics.

The difference between the figures in table 3.1 probably stems from the effort to compensate for inflation by expressing the data in a common base year. This process can only be performed in a satisfactory manner by officials with full access to the pricing system applied to products from the defence industry since items are purchased according to different price indexes.

Although one would expect both Vorobev and Barchuk to have such access, they nevertheless produced different data for 1992. One

[9] See International Institute for Strategic Studies, *The Military Balance 1993–1994* (Brassey's: London, 1993).

explanation might be that Vorobev was referring to the budget prepared by the government while Barchuk was referring to outlays.

Many prices in the Russian economy are now set by the producer based on a judgement about what the market will bear. However, the price of military equipment is set on a system-by-system basis through negotiations between the government and the producer. The MOD has 'capped' prices in an attempt to maintain its purchasing power in the face of a rapid increase in general inflation. This policy has not succeeded.

In early 1992 Vorobev explained the difficulty of preparing a budget by the uncertainty introduced by the rapid rise in prices for food, fuel and clothing.[10] In these circumstances the decision to limit prices for equipment produced to meet the Defence Order[11] proved to be of little help since the increased cost of other MOD activities squeezed the procurement budget in any case.[12] Moreover, from the perspective of the producers, the costs of their inputs—especially energy but also, to a lesser degree, labour—have risen dramatically while the prices that they can charge the MOD for finished goods have increased less quickly. This has reduced or eliminated profit margins even for enterprises which still receive orders from the MOD. Table 3.2 provides different estimates of the distribution of the defence budget between various elements.

If correct these shares indicate the increasing dominance of spending on items other than equipment within the budget. Whereas spending on equipment and R&D accounted for more than 55 per cent of the budget in 1991, by 1993 the percentage was only 25 per cent of a smaller budget. Moreover, First Deputy Minister of Defence Andrey Kokoshin has suggested that even some allocations earmarked for equipment purchases are actually being used for military construction or service pay.[13] This trend supports the public statements by officials about cuts in procurement in entire categories.

[10] See Vladykin, O., 'Interview with Lieutenant General V. Vorobyev', *Krasnaya Zvezda*, 4 Feb. 1992, p. 2, in Foreign Broadcast Information Service, *Daily Report–Central Eurasia (FBIS-SOV)*, FBIS-SOV-92-024, 5 Feb. 1992, pp. 21–24.

[11] A single Defence Order is decided which includes allocations for basic scientific research; R&D of specific weapons and dedicated technologies; the production of new armaments and military equipment; and the modernization of existing equipment. The process of drawing up the Defence Order is described in chapter 4.

[12] Vorobev offered the example of pensions, housing and 'socio-cultural construction' which together actually accounted for 70 per cent of expenditure in the first quarter of 1992. See note 5.

[13] *Krasnaya Zvezda*, 6 Oct. 1993, p. 2, in FBIS-SOV-93-194, 8 Oct. 1993, pp. 39–40.

Table 3.2. Distribution of the Soviet/Russian defence budget, 1991–93

Figures are in b. roubles.

	Soviet Union	Russia	
	1991	1992	1993
Operations and maintenance	32.1	54.7	50.0
Procurement	41.1	16.1	18.3
R&D	14.1	10.6	7.2
Other	12.7	18.6	24.5

Sources: The distributions for 1991 and 1993 are taken from the official defence budgets. The source for the 1992 figures is *Krasnaya Zvezda*, 24 Sep. 1992.

The MOD scaled back procurement dramatically in 1992. The Air Force stopped orders for MiG-29, Il-76, An-124 and An-72 aircraft, and placed only nominal orders for Tu-160, Su-27, Su-27UB, MiG-29M and MiG-29UB aircraft. In addition, only nominal orders of Mi-26 and Mi-8 helicopters were placed. New construction of warships has almost stopped while the average time for completion of ships and submarines has increased to five to nine years.[14]

Reductions in production have also taken place, reflecting reductions in procurement by the MOD (see table 3.3) as well as the continuing fall in the volume of arms exports from Russia as compared with the former Soviet Union. Unit production was still higher than the estimated aggregate value of items delivered to the MOD and foreign customers, indicating that at least for the year 1992 production without orders continued at many Russian plants (see table 3.4).

The figure of 3115.5 billion roubles presented in table 3.1 was taken from the budget submitted to Parliament by the Russian Government in May 1993. This is the figure recorded in the budget under the heading 'defence spending'. Other budget headings include allocations for defence-related activities. For example, expenditure for civil defence and payments to enterprises which must maintain a mobilization potential are contained under other budget headings. During the final round of discussions of the 1993 defence budget in July a figure of 6336 billion roubles was proposed to the Parliament by the Ministry of Finance. This figure was never approved in

[14] Comments of Vadim I. Vlasov, Assistant to the First Deputy Minister of Defence, Russian Federation, at the SIPRI workshop on The Future of the Defence Industries of Central and Eastern Europe, 29–30 Apr. 1993.

Table 3.3. Reductions in procurement by the Russian Ministry of Defence in 1992 compared with the Soviet Ministry of Defence in 1991

Figures are in percentages.

System type[a]	Percentage reduction in value of procurement
ICBMs	55
SLBMs	39
Tactical missiles	81
SAMs	80
Air-to-air missiles	80
Aircraft	80
Tanks	97
Field artillery	97
Multiple rocket launchers	76
Space satellites with space launch vehicles	34

[a] *Abbreviations*: ICBM: intercontinental ballistic missile; SLBM: submarine-launched ballistic missile; SAM: surface-to-air missile.

Source: Ministry of Defence, Russian Federation.

Parliament which regarded it as insufficient.[15] Parliament eventually approved a budget in mid-July 1993 including provision for defence of 8011 billion roubles. This figure was relatively close to the original request from the MOD (8700 billion roubles) but was not acceptable to the Russian Government (in particular, the Ministry of Finance).[16]

As a result of the stalemate between the Russian Government and Parliament the defence budget was never actually used by the MOD in regulating its relationship with the Ministry of Finance. Instead outlays were determined by a continuous process of inter-ministerial bargaining between officials from the MOD and the Ministry of Finance. The MOD consistently spent money which had not been allocated to it and subsequently sought transfers of funds to cover operating costs and to meet financial obligations to industry. As described below, this money was not always forthcoming.

[15] Radio Free Europe/Radio Liberty, *RFE/RL News Briefs*, vol. 2, no. 31 (19–23 July 1993), p. 1; Stulberg, A. N., 'The high politics of arming Russia', *RFE/RL Research Report*, vol. 2, no. 49 (10 Dec. 1993), pp. 1–8.
[16] *MIC Newsletter*, no. 8 (Aug. 1993), p. 3; *East Defence & Aerospace Update*, Aug. 1993, p. 1.

Table 3.4. Unit production by category and percentage changes in the former Soviet Union and Russia, 1990–92

	1990	1991	% change 1990–91	1992	% change 1991–92
Tanks	1 300	1 000	– 23	675	– 33
Armoured personnel carriers and infantry combat vehicles	3 600	2 100	– 42	1 100	– 48
Artillery	1 900	1 000	– 47	450	– 55
Bombers	35	30	– 14	20	– 33
Fighter aircraft	575	350	– 39	150	– 57
Attack helicopters	70	15	– 79	5	– 67
Submarines and major surface combatants	20	13	– 35	8	– 38
Strategic ballistic missiles	190	145	– 24	45	– 69

Source: *Aviation Week & Space Technology*, 28 June 1993, p. 55.

The argument between the ministries in 1992 and 1993 reflected the differences in their priorities. The Ministry of Finance saw its main priority as balancing the government budget by reducing public expenditure. Additional resource allocations to the MOD were resisted even though the purpose was to provide a social safety-net for demobilized soldiers and their families.

In the absence of a functioning budget process, the inter-ministerial disagreement has been taken to the level of the President. The fact that the MOD won some additional funds in 1992 and 1993 over the objections of the Ministry of Finance reflects its political influence. Still, some analysts maintain that the MOD has not been more successful than other powerful interest groups (notably the Ministry of Energy) in competing for supplementary allocations.[17]

Recent statements by MOD officials and defence industry representatives suggest that in spite of receiving additional funds the MOD is still unable to meet its obligations. Negotiations outside the formal process have been necessary for industry to cope with the effective breakdown in the administration of public expenditure in 1992.

[17] *RFE/RL News Briefs*, vol. 2, no. 29 (5 July 1993), p. 3; Smirnov, A., 'Hearings in the Supreme Soviet Defense Committee: military experts propose downsizing defense complex', *Kommersant-Daily*, 7 July 1993, p. 4, in FBIS-SOV-93-128, 7 July 1993, pp. 31–32; 'Money still owed to Ministry of Defense', *Krasnaya Zvezda*, 11 Sep. 1993, p. 2, in FBIS-SOV-93-176, 14 Sep. 1993, pp. 31–32.

While measures taken in August 1992 and in mid-1993 led to the payment of debts to enterprises outstanding from 1992 and the first quarter of 1993, new debts began to accumulate almost immediately in spite of the fact that the tendency towards production without orders (which continued in many locations at least through the first half of 1992) was ended in 1993. By August 1993 the debt owed by the MOD to the defence industry was around 600 billion roubles of which 271 billion was owed to design bureaus and 325 billion to serial producers.[18] In November 1993, in an effort to alleviate this situation, President Boris Yeltsin signed a decree authorizing further supplementary payments to the defence industry.[19] It included instructions that the Russian Central Bank and the Ministry of Finance must issue dedicated loans to agreed defence industry programmes.

Despite reduced production in 1993 (matching the reduction in government orders) there also appears to be a growth in 'horizontal' or inter-enterprise debt because the government has failed to make even these reduced payments to industry. Much of the lending between enterprise managers appears to have been carried out on the basis of long-standing personal relationships and was not confined only to those in Russia. Co-operation between military production enterprises located in Russia and those located in other countries of the CIS seems to have been maintained by managers on the basis of personal contacts. The same also seems to apply in the trade in raw materials and semi-finished goods.[20]

Some analysts are sceptical about the available estimates of enterprise debt. The absence of intrusive accountancy procedures has encouraged a habit of double bookkeeping and the presentation of data in ways which increase the probability of sustained state support. In these circumstances it is impossible to say with certainty whether the financial situation of the defence industry is as bad as industry

[18] *East Defence & Aerospace*, Oct. 1993, p. 3.

[19] The decree was entitled 'On the stabilization of the economic situation of enterprises and organizations of the defence industry and on measures to ensure the state Defence Order'. See *World Aerospace & Defense Intelligence*, 26 Nov. 1993, p. 7.

[20] Nina Oding, Head of Research, Leontief Centre, St Petersburg at the SIPRI workshop on The Future of the Defence Industries of Central and Eastern Europe, 29–30 Apr. 1993. According to the Russian Ministry of Defence, industry was owed around 600 b. roubles by Aug. 1993. See FBIS-SOV-93-163, 25 Aug. 1993, p. 28.

representatives claim.[21] This was also the position of former Minister of Economics Yegor Gaidar.[22]

A similar argument has been made by analysts who point to the existence of significant extra-budgetary funds to which the defence industry has access. 'The Conversion Fund', 'Research and Development Fund', 'Pension Fund' or 'Social Insurance Fund', for example, may not be used for their stated purpose but rather as a source of revenue beyond government scrutiny. In late 1991 and 1992 the Russian Government established these federal funds into which a proportion of revenues from sales tax is paid as well as contributions from industry itself. These funds—estimated to be worth nearly two-thirds of the value of the total federal budget in 1992—are said to have become 'one of the principal mechanisms enabling the old administrative structures to continue to function in a number of key sectors' including the defence industry.[23]

Reforming the budget process

It is probable that the paralysis in the budget process was a central factor in the decision by President Yeltsin to resolve the crisis in relations between the executive and the legislature. As a result of the breakdown of the present system there is pressure from the MOD to give it greater freedom to spend allocated funds. Even after a Defence Order has been agreed, the MOD still has difficulty in getting financial transfers from the Ministry of Finance.

In this reform process the preferred option of the First Deputy Minister of Defence is a US-style process of appropriation and authorization in which parliament would be able to amend a draft budget sent to it by the executive. Once a budget was authorized, the MOD would control disbursement of funds from that point on.

Ukraine

Apart from Russia, only Ukraine of the former Soviet republics has a relatively large military expenditure. Ukraine accounted for roughly 15 per cent of the total for the former Soviet Union. According to the

[21] According to Alexander Ozhegov, Chief of Economic Research Department, Russian Bank for Reconstruction and Development.
[22] Quoted in *RFE/RL News Briefs*, vol. 2, no. 43 (18–22 Oct. 1993), p. 4.
[23] Delyagin, M. and Freinkman, L., 'Extrabudgetary funds in Russian public finance', *RFE/RL Research Report*, vol. 2, no. 48 (3 Dec. 1993), pp. 49–54.

National Institute of Strategic Studies in Kiev, Ukrainian military expenditure was $1.8 billion in 1992 but fell to $367 million in 1993. However, these figures are calculated using an official exchange rate and do not account for inflation. In terms of local purchasing power these figures understate Ukrainian allocations to the military.

Approximately 100 000 non-Ukrainian servicemen returned to their country of origin during the first year of Ukrainian independence. Others have been demobilized, while about 17 000 Ukrainian troops have returned from units in other CIS members.[24] The primary expenditure is therefore to meet the costs of personnel, demobilization, housing and pension costs. Operations and maintenance (O&M) costs are also incurred by units in Ukraine. However, under the revised terms of the 1990 CFE Treaty, Ukraine inherited a very large stock of modern weapons, and expenditure for new equipment should be minimal over the next several years. Therefore it is unlikely that the Ukrainian defence industry will benefit to any major extent from government expenditure over the short term.

The Czech and Slovak republics

After the division of the Czech and Slovak Federal Republic into two independent states on 1 January 1993, all federal institutions and the former Czechoslovak armed forces were to be divided. Both countries have formed their own defence ministries.

With high unemployment, high inflation and the hesitation of foreign investors who fear political instability, even if conversion of the defence industries were seriously contemplated, the Slovak Government cannot afford the investment that is necessary to Slovakia's economic revitalization and is likely to revive the defence industry and rely more on defence production and export.

Also in the Czech Republic—as in the region as a whole—heavy reductions in the production of military goods are being implemented. Military enterprises (with the possible exception of Aero Vodochody and Tatra) are either on the verge of or in a state of bankruptcy and are forced to resort to simple survival techniques or to engage in renewed political lobbying in order to rescue their positions.[25]

[24] Presentation of Arnold Shlepakov, Ukraine, at the SIPRI workshop on The Future of the Defence Industries of Central and Eastern Europe, 29–30 Apr. 1993.
[25] 'Slovaks follow Czech industry lead', *Defense News*, vol. 8, no. 45 (15–21 Nov. 1993), p. 25.

Table 3.5. Czechoslovakia's military expenditure allocation, 1989–92

Figures are in m. korunas. Figures in italics are percentage shares.

	1989	1990	1991	1992
Personnel	9 611	7 674	8 647	10 690
	27.4	*23.8*	*31.0*	*37.5*
Operations and maintenance*a*	10 105	12 214	14 163	14 228
	28.8	*37.8*	*50.8*	*49.9*
Procurement	12 205	9 989	3 146	1 245
	34.8	*30.9*	*11.2*	*4.3*
Construction	1 812	1 346	1 235	1 680
	5.1	*4.7*	*4.4*	*5.9*
R&D	1 329	1 065	677	657
	3.8	*3.3*	*2.4*	*2.3*
Total	**35 062**	**32 288**	**27 868**	**28 500**
	100	*100*	*100*	*100*

a Includes civilian personnel cost.

Source: Compiled from the Federal Ministry of Economy and the Federal Ministry of Defence 1989–92, as well as the Federal Statistical Bureau, Prague. The 1992 military expenditure data are estimated by the Federal Ministry of Economy.

Trends in military spending

Military expenditure for Czechoslovakia was on a downward trend for the four years prior to dissolution of the country.

Table 3.5 shows the scale of the decline in the actual disbursement of funds for procurement in Czechoslovakia for the period 1989–92. While in 1989 procurement consumed 34.8 per cent of the military budget, in 1992 the share had dropped to only 4.3 per cent. The decline in the Czechoslovak defence expenditure during 1990–92 was so rapid that lack of financial resources stopped weapon development projects and led to substantial restraints in the procurement of equipment and troop training. Although military R&D fell from 3.8 per cent in 1989 to 2.3 per cent in 1992 this share is still the highest among the non-Soviet countries of Central and Eastern Europe.

According to the 1993 military budget estimates for the Czech Republic only total military expenditure will be 21.6 billion korunas ($791.1 million) which corresponds to 6 per cent of total government expenditure.[26] The Slovak Republic's defence budget for 1993 was 8.6 billion Slovak korunas ($283.3 million) equivalent to 5.4 per cent of total government expenditure.[27] The budget was described by the Defence Ministry as hardly covering the armed forces' maintenance costs, which absorb 92 per cent of the whole military budget.

Poland

The economic recovery of 1992–93 was accompanied by falling inflation rates, a surplus on the current account and a positive trade balance. Moreover, the private sector has grown substantially and Poland has become more fully integrated into the international economy during the last few years.

Nevertheless, the debt owed by state enterprises to state financial institutions and the external debt owed by the Poland (which was already $48.4 billion in 1991) have continued to grow. Perhaps the most serious indication of the Polish economy's basic weakness is that the recovery has so far been unable to redress many of the long-term structural problems.

Trends in military spending

Poland has little money to spend on defence. As shown in table 3.6, Poland has assigned 39.7 per cent of its budget for personnel costs in 1992 against 12 per cent for equipment procurement. R&D expenditure was reduced to less than 2 per cent of total expenditure in 1992. The share of the budget allocated to procurement fell dramatically after 1989, while O&M (a heading which includes civilian staff costs) has been increasing over the same period. The percentage of the budget set aside for personnel costs has fluctuated during the period. In 1993 an extra increase in personnel cost in the amount of 6.4 trillion zlotys, equivalent of 16.7 per cent of the total military budget of 38.4 trillion zlotys ($2.25 billion), was set aside for pension payments and other social expenditure.[28]

[26] Czech Republic, Ministry of Defence, *Law of Military Budgets*, C10/1993, provided by the Embassy of the Czech Republic in Stockholm, 15 Jan. 1994.

[27] Slovak Republic's Ministry of Defence, information provided by the Embassy of the Slovak Republic in Stockholm, 28 Jan. 1994.

[28] CSBM/Vienna Document 1992, *Military Budgets, Information for the Fiscal Year 1993*, CSBM/PL/93/014.

Table 3.6. Polish Ministry of Defence expenditure allocation, 1989–93

Figures are in current b. zlotys. Figures in italics are percentage shares.

	1989	1990	1991	1992	1993
Personnel	874	4 913	8 046	9 666	20 061
	39.4	*32.9*	*44.0*	*39.7*	*52.3*
Operations and maintenance*a*	676	5 034	5 441	10 060	11 992
	30.5	*33.7*	*29.7*	*41.3*	*31.3*
Procurement	502	3 312	2 813	2 926	3 942
	22.7	*22.2*	*15.4*	*12.0*	*10.3*
Construction	110	1 320	1 532	1 271	1 550
	5.2	*8.8*	*8.4*	*5.2*	*4.0*
R&D	52	366	468	451	798
	2.3	*2.4*	*2.6*	*1.9*	*2.1*
Total	**2 214**	**14 945**	**18 300**	**24 374**	**38 343**
Percentage share	*100*	*100*	*100*	*100*	*100*

a Includes civilian personnel cost.

Source: United Nations General Assembly, Reduction of Military Budgets, Military Expenditure in Standardized Form Report of the Secretary General, document no. A/4089, submitted by the Polish Ministry of National Defence, Warsaw, 7 Oct. 1992. CSBM/Vienna Document 1992, *Military Budgets, Information for the Fiscal Year 1993*, CSBM/PL/93/014.

Poland's experience is similar to that of other Central and East European countries. Under pressure for reductions in military outlays, cuts in equipment procurement are much easier to achieve than reduction in salaries, pensions and other payments to personnel.

Hungary

The effects of the domestic recession on foreign trade performance have been overshadowed by the much greater disturbance caused by the loss of the former Soviet and East European markets. While the effects of recession on import demand had been underestimated, the Hungarian Government also had to manage a substantial external debt—amounting to over $21.9 billion by the end of 1991.

Table 3.7. Hungarian military expenditure allocation, 1989–93

Figures are in b. forint. Figures in italics are percentage shares.

	1989	1990	1991	1992	1993
Operating cost^a	36.2	41.5	47.6	53.9	61.1
	75.8	*79.3*	*88.1*	*88.8*	*97.7*
Investment cost^b	11.5	10.8	6.4	46.8	3.4
	24.1	*20.7*	*11.9*	*11.2*	*5.3*
Total	**47.7**	**52.3**	**54.0**	**60.7**	**64.5**
	100	*100*	*100*	*100*	*100*

^{a} Includes O&M, personnel (military and civilian), pensions and other social expenditure.

^{b} Investment cost includes procurement, construction and R&D.

Source: Compiled from the Hungarian Federal Defence budgets for 1989–93, information provided by the Hungarian Library of Parliament, 11 Jan. 1994.

Trends in military expenditure

Economic constraints and the 1989 decision to reduce the defence budget by more than 35 per cent until 1991 significantly limit the possibilities of procuring Western equipment, although there seems to be an essential need to modernize military equipment in every sector.

As shown in table 3.7, over 80 per cent of the military budget was earmarked for operations and current expenditure during 1991–92, leaving only about 12 per cent for development and procurement while by 1993 that share had fallen to just 5.3 per cent.

The 1993 budget is intended to allow the army to maintain a 'minimum defence capability' but would have led to a deterioration in its armaments, equipment and technology in the absence of deliveries of Russian-made equipment and spare parts. However, implementation of bilateral agreements reached with Germany and with Russia in November 1992 will offset the draw-down of equipment levels to some degree. Russia will make available equipment worth $800 million in part settlement of Russia's $1.7 billion debt to Hungary.

Table 3.8. Bulgarian military expenditure allocation, 1989–93

Figures are in current m. levas. Figures in italics are percentage shares.

	1989	1990	1991	1992	1993
Investment cost[a]	1 201.2	678.3	650.1	667.5	654.4
	71.4	*41.0*	*16.3*	*12.0*	*7.6*
Operations and	481.1	980.1	3 298.0	5 103.4	8 000.1
maintenance[b]	*28.6*	*59.1*	*83.5*	*88.4*	*92.4*
Total	**1 682.2**	**1 658.1**	**3 948.1**	**5 771.0**	**8 654.5**
	100	*100*	*100*	*100*	*100*

[a] Investment cost includes procurement, construction and R&D.

[b] Operating cost includes maintenance, personnel (military and civilian) pensions and other social expenditure.

Source: CSBM/Vienna Document 1992, *Military Budgets, Information for the Fiscal Year 1993,* CSBM/BG/93/014. Bulgarian Ministry of Defence, Finance Department, provided through the Bulgarian Embassy in Stockholm, 19 Jan. 1994.

Bulgaria

Bulgaria, in contrast to Hungary and Poland, did little prior to 1990 to relax its tight central planning system or to orient its economic activities towards world market opportunities. The country has been unable to overcome the macroeconomic instability that threatens the progress made thus far and hinders foreign investment. Foreign debt—$13.5 billion of foreign debt in 1992—remained a heavy burden on the Bulgarian economy.

Trends in military expenditure

Because of the present restructuring of the Bulgarian military sector and the inadequacies of the official budget data—which are highly aggregated—the data in table 3.8 probably do not accurately represent total military outlays. Nevertheless, from table 3.8 a general observation can be made: shares of investment costs (including procurement, construction and R&D) have dropped dramatically by 63.8 per cent between 1989 and 1993. There have been substantial cuts in procurement to 5.4 per cent of the total 1993 defence budget of 8.7 billion levas ($1.18 billion).

The increase in the share of personnel costs (including military and civilian, O&M, pensions and other social expenditure) has been approximately 64 per cent during the same period. It is clear that the authorities are aware of reduced orders in the defence industries and the presence of substantial excess capacity (80–85 per cent). The arms producers are also burdened with enterprise indebtedness which has been estimated at more than 3 billion levas in 1992.[29]

Romania

Output and incomes continued to decline after the collapse of the Soviet trading and payments bloc, the impact of the UN embargo against Iraq and the former Yugoslavia, and the ineffectiveness of the government's stabilization and adjustment policies. Through 1992, gross domestic product (GDP) fell by 32 per cent and industrial output by 50 per cent.[30] The reform programme for the next four years, endorsed by the parliament in March 1993, commits the government to a reform agenda that includes reinstating consistent macroeconomic policies. More specifically, the new strategy aims to halt the decline in production, control inflation (which averaged more than 14 per cent per month in 1992) and build up the country's gold and hard currency reserves.

Trends in military expenditure

As shown in table 3.9, personnel received 37.4 per cent of military outlays for 1993, while arms procurement with about 22 per cent has a surprisingly high share of the budget. It is not known whether it is spent on acquisitions from the domestic defence industry.

IV. The dispersal of funds to industry

During the period of state socialism none of the forms of enterprise involved in the defence industry controlled either its costs or its income. This must obviously change across the region as part of the

---

[29] Engelbrekt, K., 'Bulgaria and the arms trade', *RFE/RL Research Report*, vol. 2, no. 7 (12 Feb. 1993), p. 45.

[30] *Trends in Developing Economies*, Extracts, Eastern Europe and Central Asia, vol. 1 (World Bank: Washington, DC, 1993), pp. 53–65.

Table 3.9. Romanian military expenditure allocation, 1990–93

Figures are in current m. lei, figures in italics are percentage shares.

	1990	1991	1992[a]	1993
Personnel	5 917	10 764	42 000	97 763
	17.5	*3.2*	*26.5*	*37.4*
Operations and maintenance[b]	5 749	7 704	61 068	101 437
	17.0	*23.8*	*38.5*	*38.8*
Procurement	21 151	12 807	52 901	57 570
	62.6	*39.5*	*33.4*	*22.0*
Construction	527	653	959	2 800
	1.6	*2.0*	*0.6*	*1.1*
R&D	448	450	1 590	2 060
	1.3	*1.4*	*1.0*	*0.8*
Total	**33 792**	**32 378**	**158 518**	**261 630**
	100	*100*	*100*	*100*

[a] The 1992 Submission to the United Nation gives the total figure of 138.558 b. lei; however, this does not include an additional 20 b. lei that was approved by the Parliament in July 1992, of which 5 b. lei for O&M and 15 b. lei for capital expenditure, according to Economic Committee Meeting with Co-operating Partners, Brussels, 30 Sep.–2 Oct. 1992.

[b] Includes civilian personnel cost.

Source: Laws of military budgets 1982–92, Ministry of National Defence, Bucharest, submitted through the Romanian Embassy, Stockholm, 30 Nov. 1992. For 1993: CSCE, Instrument for standardized international reporting of military expenditure, provided by the Romanian Embassy in Stockholm, 15 Jan. 1994.

effort to develop market economies. Changes in financial flows represent a fundamental change in the government–industry relationship.

In Russia, to encourage industry to continue to meet the Defence Order, producers have been allowed favourable conditions regarding value-added tax; corporate tax rates (ranging from reductions of 50 per cent to complete tax exemption); and privileged access to current and long-term credits. Plants listed as contributing to the Defence Order (described in chapter 4) are also eligible for direct investment from the Russian state budget for plant reconstruction, buying new equipment, developing manufacturing techniques and developing new materials. These concessions were part of an effort to retain the volume of production required to meet the Defence Order. Plants are also allowed to fix levels of profit sufficient to create a financial base for their industrial, scientific, technical and social development. In reality, as noted earlier, it is unclear how many producers can do this.

Enterprises in Central Europe appear to have maintained production and employment in spite of reduced military expenditure through the use of direct and indirect subsidies. Rather than government being in debt to industry—the situation in Russia—many enterprises are themselves heavily in debt to the government. In Poland, for example, in early 1993 the state was owed the equivalent of $800 billion by industry either directly to the national bank or the Ministry of Finance or to other state-owned financial institutions. The debt problem has become a major issue in government–industry relations. As one official put it:

Why should the taxpayer accept all the risks but the State abstain from any oversight and responsibility? In the state budget 4 per cent of all government expenditure has been set aside to cover the costs of bad debt owed by industry. The defence industry will have privileged access to this debt relief fund if they can present a realistic business plan. These companies will in effect be given to the banks.[31]

While trade within the WTO was carried out on a soft-currency basis, producers may now keep at least some of the proceeds of foreign defence-related sales.

In Czechoslovakia this change occurred in 1989–90 and has meant that exports are far more profitable per unit sold than they were before 1989. This creates a relationship between sales and revenue which never existed under the previous system.

In Ukraine the situation has changed more recently and less comprehensively. In the former USSR the distribution of revenues was controlled by the government. The Soviet Ministry for Aeronautical Industry collected all revenues from sales of Antonov-designed aircraft. Export revenues were collected by Aviaexport, a constituent part of the Ministry. Some revenues were distributed directly in either roubles or hard currency. The Ministry could also retain hard currency earnings and distribute roubles according to an exchange rate decided by government. Contact with customers was prohibited. This led to 'miserable amounts' of hard currency reaching the design bureau.[32]

---

[31] Jan Straus, Ministry of Foreign Economic Relations, Poland, at the SIPRI workshop on The Future of the Defence Industries of Central and Eastern Europe, 29–30 Apr. 1993.

[32] This section is based on the comments of Oleg Bogdanov, Chief Designer, Antonov Design Bureau, at the SIPRI workshop on The Future of the Defence Industries of Central and Eastern Europe, 29–30 Apr. 1993.

Antonov receives income from several sources. In addition to state funding, it receives revenues (including hard currency) from international transport operations by the Rusland commercial air-cargo carrier; revenues from the sale of Antonov-designed aircraft produced by serial production plants (the An-124, An-72, An-74, and so on); and income from repair, maintenance and upgrade of Antonov aircraft. Antonov also receives Ukrainian State Bank preferential credits in accordance with programme funding provisions.

State funding derives from the distribution of allocations from the budget title 'Program of the Ukrainian Aeronautical Industry Development' which is controlled by the newly established Integrated Ministry of Engineering, Military and Industrial Complex and Conversion. Within this ministry Antonov is in charge of issues related to aeronautical engineering. Although revenues are still paid to the government, Antonov retains more freedom of action.

V. Concluding remarks

Although there are variations in the available data, there is a consensus that military expenditure has been substantially reduced across the region. Apart from a few specific programmes by 1993, there had been an almost total curtailment of new weapon purchases. This does not reflect a lack of demand but the combination of budgetary and economic constraints and surplus stocks of CFE TLE. Since all the countries of Central and Eastern Europe will have to dispose of some TLE, further procurement in these weapon categories also raises the amount of equipment which must be destroyed to ensure CFE Treaty compliance. Nevertheless, where agreement can be reached on the conditions of transfer and where new systems can be accommodated within CFE limits, some significant new programmes have gone ahead in 1993. The most important to date have involved the restoration of arms transfer relationships with Russia by Hungary and Slovakia.

In addition to procurement, military R&D expenditure also appears to have fallen. A higher proportion of the budget is allocated to personnel costs associated with demobilization.

4. Restructuring the defence industry

Ian Anthony

I. Introduction

During the cold war the defence industries of Central and East European countries were highly integrated. In the former Soviet Union defence production and the military sector in general were seen as paramount and were not balanced against other aspects of social and economic policy to the same extent as in Western Europe or North America. Moreover, because of the political and technical control the Soviet Union exercised over other members of the WTO, the consequences of this policy were felt across Central and Eastern Europe.

Throughout the region the transformation of political relations has forced countries to devise new policies with regard to all aspects of the arms industry and its organization. The international dimensions of this transformation are considered in chapter 5. This chapter confines itself to a consideration of the national restructuring efforts undertaken in selected countries of Central and Eastern Europe. These efforts are being undertaken in three separate spheres. First, within government different agencies are having to re-define their role in the decision-making process now that the communist party no longer plays the decisive role in steering national politics. The head of the Communist Party of the Soviet Union Central Committee's Defence Department was a full member of the Politburo and often also the Minister of Defence. One of the first acts of President Boris Yeltsin upon assuming office was effectively to behead this system through a series of decrees which abolished or reduced the competence of various committees and ministries.[1] Second, government and industry are having to re-define their relationship during the transition to a market economy. Finally, within industry itself different elements within the overall production process are having to change their

---

[1] Stulberg, A. N., 'The high politics of arming Russia', Radio Free Europe/Radio Liberty, *RFE/RL Research Report*, vol. 2, no. 49 (10 Dec. 1993), pp. 1–8.

relationship towards one another and to try to form business units able to survive in a market economy.

Current information is insufficient and too fragmented to permit an in-depth analysis of these processes of restructuring, and a great deal more work will be required to explore them. Moreover, the process of restructuring is dynamic and will inevitably be affected by the outcome of developments which still remain unpredictable. The evolution of the strategic environment (with the direction of Russian policy likely to shape the policies of the countries of the region) and the economic environment in particular will shape the discussion of the defence industry.

From an intra-government perspective, sociologist Gregory Hooks has observed: 'Unlike in periods of institutional stasis, when the key issue is often one of allocating the state's resources, in periods of state building the issue is establishing an enduring administrative order and determining the locus of control in the state'.[2] In 1993 the truth of this observation has been borne out in Russia, where there has been competition between executive agencies for control over defence industrial policy-making as well as between the executive and legislative branches of government. In addition, in Russia the distribution of authority between central and regional government is also under discussion—a complication not found elsewhere in Central and Eastern Europe.

The relationship between government and industry is also changing in important ways. The measures currently being introduced include the use of negotiated contracts (rather than state orders derived from a central plan) to regulate economic transactions, decisions about the extent of private ownership to be permitted in the defence industry and new forms of government regulation—including the regulation of exports.

Finally, restructuring is taking place within industry as the linkages between the various entities involved with the production process are changing. As one observer has pointed out, 'Russian administrators and business leaders are having to come to grips with a question that has often exercised their counterparts in the West . . . namely, what are the optimal form and scale of business organization?'[3]

[2] Hooks, G., *Forging the Military–Industrial Complex* (University of Illinois Press: Urbana, Ill., 1991), p. 13.

[3] Fortescue, S., 'Organization in Russian industry: beyond decentralization', *RFE/RL Research Report*, vol. 2, no. 50 (17 Dec. 1993), p. 35.

II. Intra-government changes

The countries of Central and Eastern Europe are still in a transition period after breaking the monopoly of the communist party over both policy and administration. It is not surprising that there is a lack of clarity in the defence decision-making process.

At the risk of stating the obvious, any government must be able to resolve the conflicts between constituent parts which have different interests and responsibilities.[4] The work of defining new government structures has proceeded at different paces in different countries and no country has finalized the process. In several cases the discussion is at an early stage. The development and implementation of procurement and military industrial policies are a subordinate part of overall defence planning.

Moreover, government reform in Central and Eastern Europe is not only influenced by the need to create efficient and effective defence forces. It is also intended to reduce the political influence of the military and ensure civilian authority. This broader process of defining the role of the military within the state is a precursor to more specific decisions about arms procurement and defence industrial planning.

Under these circumstances it would be surprising if procurement decision making was functioning smoothly. As Russian MOD official Vladim I. Vlasov has observed: 'The process of equipping the Russian armed forces with arms and military equipment is taking place against the background of major military reform, the unstable internal situation in the country, the financial-economic crisis and a loss of power by Russia. In these conditions a flexible process is required'.[5]

Decision making in Russia

Russia is currently framing new structures both for military–industrial decision making and for arms procurement decision making.

---

[4] Bringing parliament into a position where it can exercise control over military expenditure and oversee the disbursement of funds was an early priority for Central and East European countries after 1989.

[5] Comments of Vadim I. Vlasov, Assistant to the First Deputy Minister of Defence, Russian Federation, at the SIPRI workshop on The Future of the Defence Industries of Central and Eastern Europe, 29–30 Apr. 1993. For a list of workshop participants, see the appendix.

Figure 4.1 identifies what appear to be the main actors influencing the Russian defence industry. It indicates that decisions by a relatively large number of different actors make some sort of impact on the industry. At the highest level the Council of Ministers of the Russian government is responsible for setting state policy. Defence industrial matters are prepared for consideration by the Council by the Department of the Military Industrial Complex and Conversion.

The extent to which the Presidential National Security Council-a recent addition to government structures-sets state policy is unclear. The creation of an Inter-departmental Committee for Scientific and Technical Problems of the Military Industrial Complex to advise the Security Council with Mikhail Malei as chairman created a potential overlap with the Council of Ministers. However, this Committee was dissolved in early 1994 and the MOD assumed the position of chief advisor to the President on defence industrial matters.

Not all government agencies wield equal influence in defence industrial policy-making. The Ministry of Foreign Economic Relations and the Ministry of Foreign Affairs play a minor role in defence industrial policy-malting through their role in the export licensing process. Both Ministries offer an opinion prior to the approval of any export licence.[6]

The central actor in determining defence industrial policy is the Ministry of Defence which is not only responsible for producing the Defence Order but has also been given a central role in deciding whether or not to allow conventional arms exports.

The Defence Order includes allocations for basic scientific research; research and development of specific weapons and dedicated technologies; the production of new armaments and military equipment and the modernisation of existing equipment. The Order is a rolling programme. The 1994 Order (finalized in August 1993) included plans for procurement through the year 2000. It also specifies the volume of production anticipated for the year ahead and how it is to be allocated among existing industrial establishments.[7] The document also identifies specific plants-around 600 of them-whose preservation is considered essential to meeting the Defence Order.

---

[6] Almquist, P. 'Arms Producers Struggle to Survive as Defense Orders Shrink', *RFE/RL Research Report*, vol. 2, no. 25 (18 June 1993).
[7] *East Defence & Aerospace Update*, Aug. 1993, pp. 1–2.

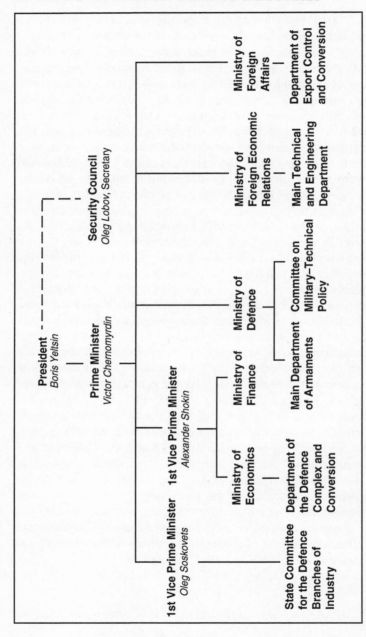

Figure 4.1. Actors influencing defence industrial decisions in Russia, 1994

The State Committee for the Defence Branches of Industry is made up of representatives of the defence complex. Originally established after the short-lived Ministry for Industry was abolished in October 1992, the Committee most of the state agencies which oversaw defence industrial activity in the Soviet Union-the State Military and Industrial Commission of the Soviet Union and eight of the nine ministries which oversaw defence production.[8] This body does not seem to have influenced the scale or direction of state orders for defence equipment. When massive cuts were made in procurement of entire equipment categories almost overnight in 1991 and 1992 industrialists were not informed of the decisions in advance let alone consulted. Subsequently they have been unable to reverse the trend to any significant degree. However, this body has been influential as a lobby arguing for continued financial support to the defence industry.

While the Ministry of Defence has developed a mechanism for determining the size and content of its budget request, no process has yet been established for adopting a state budget. As described in chapter 3, the budget process has generated tensions within the government and between the government and parliament. Within the government there is no clear process for reconciling competing budget requests from Ministries which, taken in aggregate, are higher than the amount set aside for government expenditure. Once a draft budget is sent to parliament a similar problem arises of how to accommodate parliamentary views on the volume and distribution of state spending.

In determining the state budget the Ministry of Economics, which was created in April 1991, incorporated the defence divisions of the Soviet State Planning Committee. Both this Ministry and the Ministry of Finance fall under the overall responsibility of First Deputy Prime Minister Alexander Shokhin.

The implementation of the Defence Order depends on the allocation of funds through the state budget-something which cannot be guaranteed. In arguing the case for military spending, the State Committee for the Defence Branches of Industry acts as a lobby on behalf of the defence industry.

The ninth Soviet ministry linked with the defence industry-the Ministry of Atomic Energy-survived the transition to a Russian government and remains powerful in defence of its own interests.

[8] *East Defence & Aerospace Update,* Sep. 1993, pp. 4–5.

Arms procurement decision making

The starting point for the discussion of the Defence Order is the collective body of programmes prepared by the MOD. These programmes are based on the perceived need for Russia to maintain an independent capacity to contain and repel aggression and to respond to low-intensity conflicts at the local or regional level.

The subordination of industrial policy to needs of the military in war was stressed by Defence Minister Pavel Grachev during the discussion of the Basic Law on Defence in 1992.[9] The Russian law on conversion also specifies that the needs of military doctrine ratified by the Supreme Soviet are a paramount consideration.[10]

Equipment acquisition programmes for the Russian armed forces are defined 'by the First Deputy Minister of Defence in co-operation with the Chief of the General Staff and under the control of the Minister of Defence'.[11] The Defence Minister is a member of the armed forces while the First Deputy Minister of Defence with responsibility for developing a new procurement system is a civilian. In formulating policy these individuals are supported by an Armaments Chief Directorate and a Military–Technical Policies Committee. Individual directorates specialized in particular equipment types represent a third tier below the Armaments Chief Directorate.

In determining the volume of production within the Defence Order, priority is given to the following objectives: to guarantee the capability of Russia's strategic nuclear forces; to equip mobile units; to modernize the command, control, communications and intelligence (C^3I) equipment and the electronic warfare capabilities of the armed forces; to replace advanced equipment ceded to other newly independent states on their independence; and to increase the proportion of 'smart' weapons in the inventory of the armed forces.[12] In many respects Russian planners continue to adopt the technical levels of US forces as the yardstick for measuring their own capabilities.[13]

---

[9] Grachev, P., 'Drafting a new Russian military doctrine', *Military Technology*, Feb. 1993, pp. 10–15. For a discussion of Russian military doctrine, see chapter 2.

[10] Law on Conversion of Defence Industry in the Russian Federation, 20 Mar. 1992; Section 2, Article 3.

[11] See note 5.

[12] See note 5.

[13] Fitzgerald, M. C., 'Russian military doctrine', ed. R. F. Kaufman, *The Former Soviet Union in Transition, vol. 2*, Study papers submitted to the Joint Economic Committee, US Congress (US Government Printing Office: Washington, DC, May 1993); Lepingwell, J. W.

Centre–periphery relations in Russia

In Russia the issue of centre-periphery relations has emerged as new responsibilities have devolved to the administrative regions (oblasts) and to municipal administrators in large cities with significant defence industrial dependence. One observer has suggested that the trend towards regionalism and a devolution of authority over economic decision making in Russia represent the first serious step towards reducing the dependence of Russian industry on defence-related production. According to this analysis the reallocation of labour, resources (human and material) and technology is only possible through local initiatives.[14]

Local organizations have focused on trilateral co-operation between regional government, local financial institutions and the managers of larger industrial entities in an effort to develop economic adjustment strategies.

A number of state and municipal bodies have been active in developing a regional industrial strategy in the Leningrad oblast. The Department of Conversion of Defence Industries (within the Office of St Petersburg Mayor Anatoliy Sobchak), the Deputy Commission on the Military–Industrial Complex, the regional army command, the regional Council of People's Deputies, the North-Western Regional Centre on Conversion (under the Committee of the Russian Federation on Defence Industries), and the North-West Regional Council on Conversion (under the Deputy Mayor of St Petersburg) have all expressed views on regional industrial policy.[15] Nevertheless, co-ordination has been difficult between the Leningrad oblast and the city of St Petersburg because of different political orientations. The city administration has favoured the rapid adoption of privatization measures and advocated a general exit from defence-related production. The oblast is more conservative in its orientation and is less enthusiastic about radical approaches to industrial restructuring.[16]

R., 'Restructuring the Russian military', *RFE/RL Research Report*, vol. 2, no. 25 (18 June 1993), pp. 17–24.

[14] Sergounin, A. A., 'Regional conversion in Russia: case study of Nizhniy Novgorod', *Peace and the Sciences*, June 1993.

[15] Comments of Nina Oding, Head of Research, Leontief Centre, St Petersburg, at the SIPRI workshop on The Future of the Defence Industries of Central and Eastern Europe, 29–30 Apr. 1993.

[16] Presentation of Petra Opitz, University of Oldenburg, at the Försvarets forskningsanstalt (Swedish National Defence Research Establishment, FOA) seminar on the Future of Russian Defence Industry, Stockholm, 21–22 Oct. 1993.

These local government bodies operate in regions which are of considerable size and economic importance in their own right. The Nizhniy Novgorod oblast has a population of 3.7 million and an industrial workforce of around 1.7 million people while St Petersburg has a population of 4.9 million and a workforce of around 2.4 million. In comparative terms, these regional entities represent a potential equivalent to that of Finland.

Planning at the regional level is in response to the economic crisis brought about by the collapse of orders from the centre. The scale of the crisis is illustrated in St Petersburg where, by 1993, 300 enterprises were idle, 140 enterprises were on the verge of being closed down and 400 enterprises (out of a total of 1500) were working short hours. The defence complex lost 30 per cent of scientific research workers by 1993. Moreover, St. Petersburg is not yet believed to have reached the bottom of the recession.[17]

The regions of Moscow, Nizhniy Novgorod and St Petersburg taken together probably account for 40–50 per cent of the enterprises engaged in large-scale defence-related production in Russia. Together with a few other administrative regions—Kaluga, Perm and Sverdlovsk—and republics—Bashkortostan, Tatarstan and Udmurtiya—this accounts for a large percentage of Russian defence industrial capacity.[18] Decisions taken in these regions and republics will go some way towards defining the future size and structure of the Russian defence industry.

The Governor of the oblast of Nizhniy Novgorod, the Mayor of the city of Nizhniy Novgorod and the regional Council of People's Deputies have all been active in trying to develop contacts outside Russia with a view to reducing the dependence of the local economy on production for the Russian MOD. Within the Governor's office in Nizhniy Novgorod there is now a Department of International Relations and a Department of Foreign Trade Affairs.[19] For the most part, foreign trade appears to mean maintaining or re-establishing ties with traditional industrial partners located in newly independent

[17] See note 15. A similar disruption is revealed in other regional case studies. See, for example, Kachalin, V. V., 'Defense industry conversion: a case study of the Kaluga region', *Harriman Institute Forum*, vol. 6, no. 10 (June 1993), pp. 1–12.

[18] Cooper, J., 'The Soviet Union and the successor republics: defence industries coming to terms with disunion', ed. H. Wulf, SIPRI, *Arms Industry Limited* (Oxford University Press: Oxford, 1993).

[19] Deutsche Industrie Consult, *Profile of the Region Nizhni Novgorod* (Westdeutsche Landesbank: Düsseldorf, Mar. 1993).

successor states to the Soviet Union. However, contacts have also been sought with West European partners with the specific intention of helping local industry (especially those manufacturing dual-use products) attract investment and find foreign civilian customers.

It is unclear to what extent regional bodies can implement adjustment strategies of this kind without the authorization and co-operation of the central government. In current circumstances regional bodies are able to adopt a broad interpretation of legislation such as the law on conversion and the 1991 law on privatization. Under their interpretation of these laws regional banks and industrialists have direct contact with overseas banks and customers, bypassing the central authorities. Whether this will continue to be sanctioned if and when the paralysis of central government ends cannot be known.

Decision making in Central European countries

The relative weight of economic and industrial policy questions against operational considerations appears to be different in Central Europe in comparison with Russia. Central European countries have not experienced the same level of disruption in their political and administrative processes that Ukraine has experienced. Consequently the process of administrative reform in Central Europe was certainly not easy but could be characterized as less difficult.

In terms of overall approaches to arms procurement and defence industrial decision making, in Central Europe there seems to have been less concern with the potential strategic and military implications of losing defence industrial capacities and greater emphasis on the fastest possible transition to sustainable industrial activities.

The contrast can be illustrated by the cases of Hungary and Romania. In spite of the current situation in south-east Europe and the fact that both countries border the former Yugoslavia, their policies are based on the assumption that neither country is likely to become engaged in armed conflict in the short term. The ability to produce defence-related items is subordinated to the general state of the national economy and the overall technical capabilities of industry. Military industrial policy has not been entirely decoupled from the requirements of the armed forces but the relative weight attached to procurement is lower than in Russia.

In Hungary responsibility for the defence industry and arms exports has been consolidated in one Military Industrial Office within the Ministry for Industry. This department, which is headed by military officers, is not indifferent to the needs of the armed forces. However, it is recognized that the extent to which national industries can be protected is very limited. Moreover, owing to the fact that Hungarian defence industry represents approximately two per cent of the industrial workforce, it is a less influential domestic political lobby than is the case elsewhere in Central or Eastern Europe.

At current levels of funding, orders from the MOD are not expected to contribute much to a Hungarian defence industry in which about 80 per cent of nominally domestic products were adaptations of Soviet-designed equipment. Since no new foreign customers are currently identified and no company funds are available for developing up-to-date and marketable new products, the defence industry is expected to shrink rapidly. The responsible official has stated:

the number of companies concerned with defence industrial activities is continuously decreasing as they face bankruptcy and liquidation procedures, privatization or, in more hopeful cases, division into more product-oriented units. Under these conditions research, development and manufacturing capacities have been phased out and liquidated, the most valuable specialists have left the sector and thus the accumulated special experience and knowledge is being reduced to naught.[20]

The Hungarian Government intends to preserve some capacities including, where necessary, giving subsidies. These capacities are those concerned with designing, developing and manufacturing telecommunications systems and networks as well as soft-skinned vehicles, and biological, medical, chemical and radiation protection devices and instruments. These are areas where the Hungarian Government has decided that there may be long-term possibilities for commercially viable production. These sectors may be assisted during the next few years, which is seen as a transition period.

In Romania, 'the restructuring of the defence industrial base is regarded as part of the overall industrial restructuring process. The institutions responsible for designing and implementing the government's economic reform programme have assessed the defence industrial base and defined the restructuring strategy'.[21] Three alternative approaches have been discussed: increasing the funding of military R&D through the defence budget; consolidating production into government owned arsenals focused exclusively on arms production; and developing international co-operation in production.

One conclusion is that military expenditure should be concentrated on R&D on dual-use technologies, of which the most important relates to computers. The creation of private companies or mixed companies with government and private ownership was preferred to an arsenal-type arrangement. Finally, government procurement would be competitive with cost considerations paramount given the low level of funding for defence.

Thus, a shrinkage in the industries producing military-related items is expected. Nevertheless, production of consumable items will continue in order to maintain limited capacities in 'key defence sectors' (such as ordnance) and 'a robust maintenance and repair capability' (spare parts for dedicated military systems).[22]

III. Changes in government–industry relations

As the countries of Central and Eastern Europe move towards market economies, it is necessary to move from resource allocation through a centrally formulated plan to a situation in which economic trans-actions between parties are regulated by contract. It is also necessary to resolve the question of ownership of industry. Should none, some or all of the state-owned enterprises be privatized? As the ultimate purpose of the defence industry is to contribute to military operations, even if production facilities act as autonomous economic units in private ownership, it is reasonable to assume that governments will insist on overseeing their activities and that they will regulate exports by the defence industries.

The introduction of contracts

In Russia preparations for the introduction of contracts to regulate relations between the MOD and manufacturing plants were complete

---

[21] Maftei Rosca, Director General, Ministry of Industry, Romania, at the SIPRI workshop on The Future of the Defence Industries of Central and Eastern Europe, 29–30 Apr. 1993.
[22] See note 21.

in April 1993.[23] However, the deadlock between the executive and the legislature meant that the passage of the law 'On State Defence Orders' was blocked.[24] Standard contracts for defence R&D and the purchase of military equipment are being developed by the MOD in the framework of the Russian law 'On Deliveries of Products and Goods for the State'. These contracts are intended to include a schedule in which work will be completed, and protocols of agreement on prices, the dispensation of funds, compensation for default and adjustments to the contract price. Under the contracts the responsibility rests with the prime contractor to reach and manage agreements with subcontractors about the development and production of necessary equipment.

Discussions between government and industry still refer to an index of contract prices expressed relative to past-year, rather than current, rouble prices. This reflects the impact of Russian inflation. As noted in chapter 3, the payment arrangements between the government and industry are not functioning smoothly. However, even if price stability can be achieved it is unlikely that fixed contract arrangements can be made to apply across the lifetime of a major defence programme. The nature of the procurement process means that a high degree of government–industry co-operation is required in the management of programmes and contracts reflect this.

Murray Weidenbaum, former Chairman of the Council of Economic Advisers to US President Ronald Reagan, has observed that

. . . it is generally impossible to predict the cost, schedule, performance or quantity of the final product with enough precision to permit the buyer and seller to write a firm contract covering the entire process. Instead the two parties establish an uneasy alliance sharing risks and management responsibilities under the aegis of a contract that is at times little more than a baseline for negotiations over numerous changes in the course of a programme.[25]

Ownership of the defence industry

State ownership of the defence industry is common. In the USA— where most defence industrial facilities are privately owned—

[23] See note 5.
[24] *RFE/RL News Briefs*, vol. 2, no. 29 (5 July 1993), p. 3.
[25] Weidenbaum, M., *Small Wars, Big Defense: Paying for the Military after the Cold War* (Oxford University Press: New York, 1992), p. 133.

government-owned factories nevertheless compete directly with private suppliers.[26] In addition, some companies in the USA and the UK are 'GOCO' (government-owned, contractor-operated) with private management and administration but state-owned land, buildings and machinery. Moreover, in the USA and the UK the armed forces retain significant industrial capacity within units responsible for repair and maintenance.[27] In other countries with market economies—France, Israel, Italy and South Africa—most or all of the defence industry is publicly owned, although in all of these countries privatization is either under way or being considered.

The ultimate objective of governments in Central and Eastern Europe is the creation of market economies. Additional objectives include reducing the power of the state to exercise control and influence over enterprises; the creation of a class of managers who will run enterprises as commercial businesses; the introduction of private property ownership; increased efficiency in the utilization of resources by enterprises; and generating revenues for the government through the sale of assets.[28] If decisions are taken which result in major job losses in Central and East European defence industries, governments may well want to be insulated from direct responsibility for such decisions.

Many industrialists in Central and Eastern Europe are inherently predisposed to achieving privatization and diminishing state regulation as a response to their experience with state socialism. Some argue that if privatization is blocked, producers will choose to stop defence-related production. Nevertheless, even after the introduction of elements of a market economy, defence industries may be intrinsically unsuitable for privatization.

In Russia and Ukraine the privatization process is at an earlier stage than the programmes being pursued in Central Europe. Whereas it is clear that in Central European countries privatization will mean a genuine change of ownership and control, the same cannot be said for Russia. The State Programme for Privatization adopted in June 1992 divided industrial enterprises in Russia into five groups. All defence-

[26] In shipbuilding, this has been less the case after 1983, when new construction was concentrated in private yards. Whitehurst, C. H., *The US Shipbuilding Industry: Past, Present and Future* (Naval Institute Press: Annapolis, Md., 1986). However, it remains true in the land system sector, especially as regards artillery.

[27] *Aviation Week & Space Technology*, 21 June 1993, pp. 69–71.

[28] *Legal Aspects of Privatization in Industry*, United Nations Economic Commission for Europe, ECE/TRADE/180 (UN: New York, 1992), chapter 1.

related enterprises are located in one of three groups: (*a*) the patent service, standardization and metrology agencies and machine testing stations; (*b*) entities working with nuclear materials or space; and (*c*) entities repairing or manufacturing any weapon system or component, munitions, explosives and pyrotechnic products and related design and R&D organizations.

Enterprises within the first two groups are excluded from privatization. Entities within the third group may be privatized through a special decision by the national government or the governments of the republics in the Russian Federation. However, this process can perhaps best be described as the clarification of state property rights rather than privatization. If permission is granted, the Russian Government has the right to keep 51 per cent of the shares for the first three years of operation. There are many cases where several government agencies all insist on their exclusive right to control particular industrial plants. The process of 'privatization' with 51 per cent state ownership will clarify which government agencies have operational control over which assets. This may prepare the ground for a second round of more genuine privatization at an unspecified future date.[29]

Some Russian analysts believe that enterprises deriving more than 20–30 per cent of their revenues from defence contracts cannot be privatized in the true sense of the word regardless of government intentions (although it may be possible for them to sell some assets) because they are unattractive to investors.[30] Others suggest that whether or not investors could be found, the managements of many defence enterprises are hostile to privatization as they would then lose access to government subsidies. These managers prefer a process which gives them greater autonomy within the existing ownership structure.

As the defence industry is not a homogeneous industrial sector it is not surprising that attitudes to privatization are influenced by levels of dependence on MOD orders. The geographic location of industry is also an important factor influencing attitudes. Most of the examples of enterprises which have made public statements in favour of privatization have been in 'dual-use' industrial sectors (mainly electronics and

[29] Yevgeniy Kuznetsov, Centre for Economic and Mathematical Studies, Moscow, at the FOA seminar on the Future of Russian Defence Industry, Stockholm, 21–22 Oct. 1993.

[30] Khroutskiy, V., Koulik, S. and Ushanov, Y., *Russian Military Industry: Present Realities and Conversion Efforts* (Center for Conversion and Privatization: Moscow, 1993), pp. 24–25.

shipbuilding) and located around the periphery of Russia—either close to Finland or in the far east. This suggests that privatization is seen as a form of industrial restructuring which can contribute to helping an enterprise reduce its dependence on the defence sector or even to stop doing business with military customers entirely. This is another reason to believe that defence industries in Russia will tend to remain in state ownership and to be concentrated in more central regions that lack economic alternatives.

The same seems to apply in Ukraine where Leonid Kuchma (Prime Minister between October 1992 and September 1993) favoured privatization in the service sector, light industry and agriculture but stated that nuclear, energy-related and military industries must remain in state ownership. Kuchma was formerly director of the Southern Machine Construction Plant in Dnepropetrovsk, the largest rocket and missile production plant in the former Soviet Union.[31]

In Central Europe businesses considered part of the defence industry but with relatively little defence-related activity are likely to be the first to be transferred to private ownership. In Hungary and Romania, enterprises of this kind make up the bulk of the industries defined as part of the defence industrial base.

In Romania, 75 per cent of the defence industrial base consists of entities which are integrated with some kind of civilian activity. Not only are management and administrative staff integrated, but military and civil producers share access to energy and machinery. This is true across the range of R&D, production and repair activities.[32]

In Hungary the defence industry formally includes those producers whose military-related output represents 10 per cent or more of total output. This involves around 70 enterprises of which only five have more than 50 per cent military-related output and only one with exclusively military-related production. The 20–25 enterprises considered most important for defence production are under strict supervision with managers appointed by and reporting to the Minister of Industry. However, other enterprises (such as some chemical manufacturers) with little regular defence-related production apparently receive state subsidies to maintain idle capacity for use in military production.[33]

---

[31] Solchanyk, R., 'Ukraine: the politics of reform', *RFE/RL Research Report*, 20 Nov. 1992, p. 4.

[32] See note 21.

[33] Judit Kiss at the SIPRI workshop on The Future of the Defence Industries of Central and Eastern Europe, 29–30 Apr. 1993.

As these enterprises are privatized, their relationship with government is likely to change as governments consider whether or not to continue paying subsidies to companies nominally operating under commercial conditions in order to retain idle capacity. This decision will be determined partly by the fact that those enterprises which find it the hardest to adapt to commercial operations are often also those considered essential to national defence. While product sectors which were the easiest to modify have found that engineers and specialists have already left (especially the younger ones) this has not happened for those engaged in artillery and small arms production.[34]

Government regulation of defence industries

The tension between the desire of industrialists for freedom of action and the government requirement for control over military-related aspects of national life finds expression in government regulation of the defence industry. It is a misperception that in market economies the defence industry is unregulated. Regulations can have many functions: to ensure supplies of critical products or materials; to ensure quality control in defence products; to permit oversight and audit of government expenditure; and to oversee the export and import of military-related goods and technologies. All forms of regulation impose costs on industry but none are likely to be abandoned completely since each fulfils a legitimate function.

Retention of mobilization capacities

Retaining industrial capabilities to support the armed forces requires government intervention since the natural tendency for commercial companies is to shed any excess capacity. In Russia this aspect of government intervention reveals the incompatibility of the goals which have been set for the defence industry: to operate according to commercial principles within a regulatory environment which mandates the retention of unproductive capacities.

The complexity of modern weapons effectively rules out 'surge production' of the type undertaken during World War II,[35] and even

[34] See note 20.

[35] The unsustainability of surge production has been recognized since the 1960s. See, for example, Raymond Garthoff in his introduction to Sokolovsky, V. D., *Military Strategy: Soviet Doctrine and Concepts* (Praeger: New York, 1963).

the USA depends on adequate stockpiles and an effective division of labour within a stable alliance for secure supplies.[36] Nevertheless, all governments retain the legal power to intervene in industrial decisions.[37]

In 1991 the Deputy Chairman of the Russian State Defence Committee observed that the capacity for wartime mobilization of industrial assets is 'a sacred cow, the principles of which have remained unchanged since the 1930s'.[38] In 1992 the law on conversion passed by the Supreme Soviet of the Russian Federation also included this approach.[39]

The law stated that 'on the basis of agreements and using defence-allocated funds, conversion enterprises will assure creation, maintenance and development of mobilization capacities'.[40] For this reason the law states that 'enterprises and structural units of enterprises for mobilization purposes not used in current production are not subject to privatization'.[41] In general the law on conversion is at least as concerned to limit any reduction in industrial capacity as it is to facilitate a reorientation of production.

In July 1992 President Yeltsin confirmed this in a decree stating that any enterprise which chose not to bid for government contracts but which was formerly a supplier of the MOD should retain production capacities. Any manager failing to maintain adequate capacities would be criminally liable. Specific legislation addressing this issue was in preparation in 1993.[42]

---

[36] Taft, W. H., Abshire, D. M., Burt, R. R., Merrill, P. and Woolsey, R. J., *Transatlantic Defense Co-operation in a Time of Transition* (Center for Strategic and International Studies: Washington, DC, 1993).

[37] In the USA the Defense Production Act and the National Defense Commerce Act give the US President wide powers of intervention in industry including the power to offer direct subsidies to companies and instruct them to fulfil military contracts in preference to civilian ones. These powers were invoked not only in wartime but also were used in the 1980s when problems emerged with companies which were the sole source of critical materials for NASA or the Department of Defense. See Linke, S. R., *Managing Crises in the Defense Industry: The Pepcon and Avtex Cases*, McNair Paper no. 9 (Institute for National Strategic Studies, National Defense University: Washington, DC, July 1990).

[38] Quoted in Cooper, J., 'Conversion of the defence industry in the CIS: the issue of mobilization capability', presentation to the NATO–CEE Conversion Seminar, Brussels, 20–22 May 1992.

[39] See note 10.

[40] See note 10, Section 1, Article 2.4.

[41] See note 10; Section 2, Article 6.5.

[42] Julian Cooper refers to a Draft Law on Mobilization Preparation and Mobilization in the Russian Federation. See Cooper, J., *The Conversion of the Former Soviet Defence Industry* (Royal Institute for International Affairs: London, 1993), p. 29.

In Russia there is a suspicion that legislation will not be seen as a necessary last resort but instead will be used to justify regular political intervention in industry. This led Vladimir Alferov, Executive Director of the League of Assistance for Defence Enterprises, to observe that 'all the directives from above are driving our directors mad: increase the output of civilian goods but do not decimate military production. What on earth does this mean?'[43]

In Central Europe the primary focus of government intervention in industry is likely to be through export licensing.

Export regulations

During the period of state socialism, the export of military equipment was under tight political control. The regulatory system has been changed and replaced by a system in which any exporter making or selling items on official control lists should obtain a licence prior to delivering controlled goods to a foreign customer. However, authorities have found it difficult to convince producers—many of which do not think of themselves as part of the defence industry—of the need for intrusive administrative regulations. These regulations carry associations with the old socialist system.[44]

In formulating new export control regimes, Central and East European countries have put most of their energy into the regulation of items which may contribute to the spread of weapons of mass destruction. In most cases the control lists in use in Central and Eastern Europe for licensing purposes have been based on the Industrial List and the International Atomic Energy List developed by the Co-ordinating Committee on Multilateral Export Controls (COCOM). They have also been revised in accordance with the control lists used by the Australia Group, the Nuclear Supplier Group and the Missile Technology Control Regime (MTCR).

In Hungary it has been difficult to communicate licensing requirements to all of the relevant manufacturers. There are no national, sectoral associations equivalent to bodies such as the Aerospace Industries Association in the USA or the Defence Manufacturers

[43] *Kommersant*, no. 15 (12–18 Apr. 1993).

[44] János Csendes, Ministry of International Economic Relations, Hungary, at the SIPRI workshop on The Future of the Defence Industries of Central and Eastern Europe, 29–30 Apr. 1993.

Association in the UK which can be used by the government to communicate with industry.

The introduction of regulations on exports of conventional weapons has been slow because of the difficulty of framing legislation, the reluctance of industry to accept strict regulations and the lack of international pressure for export controls on these weapons.

In Poland the lack of understanding within Polish industry and among foreign journalists of how the new regulations function has led to several 'scandals'. In each case individuals from Poland took advantage of their new freedom to make contact with potential foreign customers. While such discussions are legal, no foreign sales can be concluded without the permission of the Polish licensing authority (the Central Engineering Board, CENZIN, within the Ministry of Foreign Economic Relations).

Several Central European countries—including Bulgaria, the Czech Republic, Hungary, Poland and Romania—have drafted or enacted laws governing exports. This legislation has been closely modelled on Western practice, and Germany, Sweden, the UK and the USA have all offered advice and training. The regulations share the following features:

1. Exporters should seek government permission before making foreign sales of controlled items.

2. Permission is likely to be granted for conventional defence equipment except where there is a consensus that the buyer represents a threat to international stability. This is usually taken to mean that the potential recipient is under mandatory United Nations embargo, a group which includes governments such as those of Iraq, Libya, and all of the states to emerge on the territory of the former Yugoslavia which were important customers for Central and East European suppliers.

3. Permission will be denied for items that contribute to the development or production of weapons of mass destruction (including ballistic missiles).

There was an initial willingness among Central European countries to discuss export limitations in 1990 and 1991. The interest was conditional on compensation for lost export sales through increased access to West European civilian markets and financial assistance for industrial restructuring.[45] Neither form of compensation has been forthcoming and interest in conventional arms export control is now low across Central and Eastern Europe. On the contrary, advocacy of export control for conventional weapons is now seen as a device used by major North American and West European governments to secure competitive advantages for their own industries.

Under these circumstances it seems even less likely that this form of government regulation will be a significant barrier to industrial activity. It is more likely that industry and government in Central and Eastern Europe will develop a close partnership of the kind which exists in most market economies with significant defence industries. In these countries close engagement in the marketing of industrial products (including defence equipment) has become a feature of overseas travel by senior politicians.[46]

Russia

In the Soviet Union arms exports were always under close executive control.[47] After the dissolution of the Soviet Union Russia moved quickly to create a new administrative export control apparatus in 1992. The power to control exports was established by presidential decree on 22 February 1992, banning unauthorized trade in 'strategic goods and commodities' (including precious stones and metals as well as defence materials and equipment).[48]

In a further decree on 12 May 1992 President Yeltsin revealed the procedures for control of exports and imports of arms, military equipment, construction and services. Controlled items require a licence issued by the Ministry of Foreign Economic Relations. The original list of controlled items was inherited from the Soviet Union. However, the export control apparatus of the Russian Federation has subsequently been expanded to include four additional control lists. The five lists now in use consist of the munitions list inherited from the Soviet Union; a list of materials, equipment, technologies and

[45] Cupitt, R. T., 'The political economy of arms exports in post-communist societies: the cases of Poland and the CSFR', *Communist and Post-Communist Studies*, vol. 26, no. 1 (Mar. 1993), pp. 87–103.

[46] This issue is revisited in chapter 6.

[47] For a description, see Wulf, H., 'The Union of Soviet Socialist Republics', ed. I. Anthony, SIPRI, *Arms Export Regulations* (Oxford University Press: Oxford, 1991).

[48] Sergounin, A. A. and Subotin, S. V., *New Russian Trade Policy: Needs and Opportunities*, Working Paper no. 15 (Centre for Peace and Conflict Research: Copenhagen, 1993), pp. 6–12.

scientific research used for producing conventional arms; a list of chemicals and technologies designed for peaceful purposes but which can be used for chemical weapon production; a list of dual-use equipment and appropriate technologies for nuclear purposes; and a list of equipment, materials and technologies used for missile production.[49]

Decisions on licence approval and export policy are made by the Commission for Military Technical and Economic Co-operation (KVTS) headed by the vice-prime minister with the Ministers of Foreign Economic Relations, Foreign Affairs, Defence, Economics, Industry, Finance and Security as well as the Chairman of the State Committee for State Property and the Head of the External Intelligence Service as members. This Commission is similar to and inherited many of the functions of the Soviet Defence Council as reconstituted by President Gorbachev in 1989.[50]

In 1994 export regulations remain based on presidential decrees rather than a law. In 1993 an inter-agency group led by representatives of the Ministry of Foreign Affairs drew up a draft basic law on import and export regulation.[51] In the previous Russian Parliament, a group drawn from four permanent Committees—on Industry and Energy; International Affairs; Defence and Security; and Budget, Taxation and Pricing—had worked on an export control law for more than 18 months but this group had not yet been presented to Parliament when it was dissolved on 21 September 1993.

In 1993 the export control procedures were modified to limit the number of Russian business entities licensed to carry out foreign sales of controlled goods. In the original apparatus established in May 1992 foreign trade corporations including the producers themselves were permitted to initiate independent contacts with prospective foreign customers.

This led to a series of cases in which the foreign contacts established by some producers and dealers were embarrassing to the Russian Government. Several producers and dealers were trying to manage the same transaction simultaneously and without co-ordination.[52] As a result the Government limited the number of

[49] Correspondence with Nikolai Revenko, Counsellor for Disarmament and Military Technology Control, Russian Ministry of Foreign Affairs, 9 Nov. 1993.

[50] See note 47, p. 171.

[51] Kortunov, S., 'The Russian perspective', eds K. Peabody O'Brien and H. Cato, *The Arms Trade in a Transitional Economy* (Global Outlook: Palo Alto, Calif., Oct. 1993).

[52] 'Russian defence sales: the insider's view', *Military Technology*, Dec. 1993, pp. 40–57.

business entities licensed to initiate foreign contacts to three government-owned agencies—Oboronexport, Spetsvneshtekhnika and the GUSK—and Promexport, an organ of the State Committee on Defence Industries. Although they increasingly market their activities through the commercial press, each of these agencies is subordinate to the Main Directorate for Military–Technical Co-operation within the Ministry for Foreign Economic Relations and can trace its lineage to corresponding structures within the bureaucracy of the former Soviet Union. Oboronexport was formerly the General Engineering Department in the Soviet Ministry of Foreign Economic Relations; Spetsvneshtekhnika was formerly the General Technical Department and GUSK was the General Co-operation Department.[53]

In November 1993, President Yeltsin furthered modified the management of arms exports by placing all three bodies under the overall control of a single entity, the Rosvooruzheniye.[54]

The Czech Republic

In 1990 and 1991 the Government of the former Czechoslovakia took two decisions which formed the framework for export regulations. First came Federal Government Act 256/1990 of 4 May 1990 'on determining exports and imports of goods and other foreign trade activities requiring a licence' followed by the parliamentary decision of 21 March 1991 which required all arms sales to have a licence from the Ministry of Foreign Trade.[55] In the Czech Republic the licensing authority is the Ministry of Industry and Trade.

In 1992 a decree issued by the Czechoslovak Ministry of Foreign Trade expanded the range of controlled goods and technologies by introducing a list based on the COCOM industrial core list, a list of nuclear dual-use items and a list of chemical weapon precursors.[56] These lists are still in use in the Czech Republic. However, in 1993 the control lists were being amended further to incorporate items on the MTCR Equipment and Technology Annex.

---

[53] 'Russian arms export policy detailed', *Military Technology*, Oct. 1992.
[54] *Moscow News*, no. 5, 4–10 Feb. 1994, p. 3.
[55] Matousek, J., 'Czechoslovakia', in Anthony (note 47).
[56] *The Worldwide Guide to Export Controls, 1992–93* (Export Control Publications: Chertsey, Surrey, UK, Nov. 1993).

Hungary

The Hungarian defence industry has become highly export dependent. In 1991 and 1992 Hungary processed 3000 licence applications for goods worth $500 million, and the government is examining the possibility of deregulation to reduce the burden that maintaining such strict export regulations imposes on both government and industry.[57]

The majority of security export controls in Hungary are applied to dual-use products as the arms industry which, never large, has further diminished in the past few years.

Hungary appears to be the most advanced of the Central European countries in developing a liaison between the export control authority, industry and the customs service which is a pre-requisite for the effective implementation of export regulations. In part this reflects the investment made in Hungary in the context of enforcing the United Nations mandatory arms embargo against the former Yugoslavia and trade sanctions against Serbia and Montenegro.

Poland

In common with other Central European countries, Poland has introduced a licensing system to regulate exports of sensitive items.[58] CENZIN, the licensing body, is a department within the Ministry of Foreign Trade, although individuals from other interested authorities are seconded to the department. In the case of Poland the primary regulations were introduced in 1989 in the form of government orders issued by the Minister for Foreign Economic Co-operation.[59]

In 1990 as part of a wider discussion of restructuring the defence industry it was decided to try introducing competition and to widen the use of commercial practices in the defence industry. Producers were given greater freedom to establish independent contacts with prospective foreign suppliers. Government regulation was exercised through the requirement to obtain a licence before delivering any goods. This system led to a series of scandals related to export controls which caused political embarrassment both within Poland and

[57] See note 44.
[58] Zukrowska, K., *The Dilemmas of Polish Arms Industries in the Period of Systemic Change*, Materialen und Dokumente zur Friedens und Konfliktforschung Nr. 9 (Berghof Stiftung für Konfliktforschung: Berlin, 1992).
[59] Zukrowska, K., 'Poland', in Anthony (note 47).

between Poland and other countries—notably Germany and the USA.[60]

In 1992 Minister for Foreign Trade Adam Glapinski proposed new regulations aimed at placing further restrictions on the extent to which business units licensed to manufacture arms can establish contact with prospective foreign customers.[61]

Slovakia

In Slovakia the acts drawn up by the legal department of the Federal Ministry of Foreign Trade of the former Czechoslovakia still form the basis for export control. The implementation of export control measures has been complicated by the judicial separation into two sovereign states and by statements made by the new government suggesting that export control is a low priority.[62]

In terms of marketing and distribution, Slovakia has established its own state-owned trading bodies based on Slovak employees from the armed forces and OMNIPOL, the trading firm run by the former Czechoslovakia.[63]

IV. Changes in intra-enterprise relations

As noted in the introductory chapter there is no legal equivalent in Central and Eastern Europe of the firm or corporation found in market economies. Moreover, in several important respects not only the individual business units but also the relationships between them differ from those in market economies. While efforts are being made to restructure these business units, comparatively little progress has been made.

A study of the economics of socialism from as early as 1972 stated that the higher stages of socialist development required 'decentralized decision-making and a greater concern for the consumer pref-

---

[60] The background to several of these scandals is outlined in Sabbat-Swidlicka, A., 'Poland's arms trade faces new conditions', *RFE/RL Research Report*, vol. 2, no. 6 (5 Feb. 1993), pp. 49–53.

[61] *Jane's Defence Weekly*, 11 Apr. 1992, p. 599.

[62] *RFE/RL Research Report*, vol. 1, no. 47 (27 Nov. 1992), pp. 58–9. This issue is revisited in chapter 6.

[63] Laurent, P. H., 'Czech and Slovak arms sales policy: change and continuity', *Arms Control*, vol. 14, no. 2 (Aug. 1993), p. 157.

erence'.[64] Nevertheless, by the late 1980s decentralized decision making was unknown in the defence sector. Until very recently the government determined costs and controlled revenues as well as directing product development, setting production levels and absorbing all of the output of the industry, either for the armed forces or for export. Managers needed no expertise in investment, marketing or distribution. Far more important was a thorough knowledge of the inner workings of the relevant ministries and central planning organizations. Tasks which in a market economy would usually be performed by an integrated company or industrial group were separated and co-ordinated by government agencies.

Current patterns of organization inhibit the process of change for those businesses which are trying to diversify their activities or leave the defence sector altogether. A US executive, describing his experience of discussions and visits to Russia in 1993, observed:

The new industrial fabric of East European countries should be built using as many parts as possible of existing production units. However, conversion to a market economy requires more. Businesses will have to be developed. This is not just a matter of modernizing the equipment in existing production units and converting them from one product line to another. A successful business includes many disciplines: market research, technical research and development, manufacturing, distribution, marketing, sales, customer services, administration, accounting and cost control etc. The reality in most of the industrial complex units built in eastern Europe in the past is that vital components of such a system are missing.[65]

In fact these disciplines are less vital to a military than a civilian business. In the defence sector the terms 'market research', 'marketing' and 'sales' all mean in effect the same thing—watching developments in the defence budget and negotiating with the MOD. Technical R&D is often conducted under the direction of government agencies while 'customer services' and 'distribution' in practice mean liaison with the armed forces.

[64] Wilczynski, J., *The Economics of Socialism* (George Allen & Unwin: London, 1972), p. 171.
[65] Statement by Ursus Jaeggi, Director of Governmental Affairs, Du Pont International before the Joint Hearings of the Committee on Foreign Affairs and Security, Committee on Budgets; Committee on Economic and Monetary Affairs and Industrial Policy, Committee on External Economic Relations, European Parliament, 28 Apr. 1993.

The central question which managers must answer and which will determine whether and how they need to restructure their operations is whether or not to remain in the defence sector.

The organization of business units in Central and Eastern Europe

In Central and Eastern Europe a variety of more or less free-standing entities performing functions related directly to manufacturing can be identified. Strictly speaking none of these should be called a company. As William Butler has observed, the word company is a misnomer and, until the dissolution of the system of state socialism, the word was never used, being reserved for entities in capitalist countries.[66] Across Central and Eastern Europe the defence sector remains dominated by state-owned enterprises of different kinds. The Russian law on conversion identifies design or scientific research organizations, science and production associations, production associations and plants.

Design or scientific research organizations with some production capacity (sufficient for prototyping and advanced development) but without serial production capacities seem to have been largely confined to Russia though there are isolated examples elsewhere in Central and Eastern Europe (such as Antonov in Ukraine).

Science and production associations have the capacity to perform a wide spectrum of operations within the overall production cycle. They may be very significant industrial assets with tens of thousands of workers distributed across five or more locations. While few if any such associations undertake all basic research and technical development internally, they have significant in-house capabilities in this regard. However, most of these capacities would presumably be directed at developing processing technologies. These also seem to have been largely confined to Russia although there are isolated examples of such organizations in Central European countries—PZL in Poland, for example, or ZTS in Slovakia.

Production associations group a number of factories performing serial production or closely interdependent manufacturing tasks. Plants or single factories are engaged only in one limited manufacturing task. These structures are not unknown in market economies

---

[66] Butler, W. E., *Companies and Contracts in Russia and the CIS* (Royal Institute of International Affairs: London, 1993), p. 4.

where defence producers are often insulated from the civilian manufacturing sector even within companies which on the face of it seem to be diversified. Jacques Gansler has underlined this in his study of the US defence industry. He notes: 'it is critically important to distinguish between number of firms and number of plants. Both are important, but in the defence area there is a tendency for a plant to be the equivalent of a firm—to have its own engineering, marketing, management and so on'.[67] Nevertheless, some unique features of industrial organization in Central and Eastern Europe exist.

Perhaps the clearest distinction between the industrial units in Central and Eastern Europe and counterparts in defence industries elsewhere is the important role enterprises play in meeting the social needs of the labour force. Functions undertaken in the private sector or by government in market economies—such as health care, child care, and the provision of sports and recreation facilities—are often organized at the enterprise level in Central and Eastern Europe.

Another unique feature is the relationship between design bureaus and serial production plants. The US Government is heavily engaged in military research not only in terms of providing funds but also in the more practical sense of running its own laboratories and establishments such as the Naval Weapons Center at China Lake, California. In the UK the Defence Research Agencies also provide specialist support to industry. Companies faced with specific technical problems that cannot be solved in-house can subcontract with the Defence Research Agencies in an effort to find a solution. Nevertheless, for the most part research and especially product development take place within single companies. These companies expect to benefit by having the preference if not an exclusive right to produce equipment that they have developed successfully.

This was not the case for complex systems within the Soviet production system where the tasks of product development and serial production were separated. Several designers could share the same production associations according to the decision of government planners. Table 4.1 gives an example of the distribution of production for fighter aircraft developed by the design bureaus Mikoyan and Sukhoi. The design bureaus are concentrations of highly educated and skilled workers while the majority of employment is concentrated in the production associations.

---

[67] Gansler, J., *The Defense Industry* (MIT Press: Cambridge, Mass., 1982), p. 45.

Table 4.1. Distribution of production of fighter aircraft designed by Mikoyan and Sukhoi, 1993

	Design bureau	
Location	Mikoyan	Sukhoi
Moscow	MiG-21	
	MiG-23	
	MiG-29	
Nizhniy Novgorod	MiG-25	
	MiG-29	
	MiG-31	
Novosibirsk		Su-27
Irkutsk	MiG-27	Su-27
Ulan-Ude	MiG-27	Su-25
Komsomolsk-Na-Amure		Su-27
Tblisi (Georgia)	MiG-21	Su-25

Source: Presentation of Alexander Ozhegov at the FOA symposium on the Russian Defence Industry, Stockholm, 20–22 October 1993.

As indicated in table 4.1, some plants produce designs exclusively from one or another design bureau while others produce designs from both. The decision about what is produced is taken by the Government and not by the design bureau or the production association. This separation of product development from serial production in separate and independent units with the government brokering relations between them makes it unlikely that either form of unit could survive in the absence of close Government involvement.

Another form of industrial organization that is apparently unique to centrally-planned economies is the co-located production associations combining many different types of industrial activity—ranging from the extraction and processing of raw materials to construction.

The phenomenon of co-location has made the available data on the structure of enterprises in the defence industry difficult to interpret. Production in Russia seems to be undertaken by very large units. Table 4.2 illustrates this structure and suggests that the defence industry is highly concentrated. However, as studies of specific regional industries have been carried out it is clear that within each of the production units there is typically a wide variety of forms of activity under way that have little or nothing directly to do with the produc-

Table 4.2. Employment structure of defence industry enterprises, 1993

Number of employees	< 1 000	1 001–5 000	5 001–10 000	> 10 000
Percentage of defence enterprises	5.8	49.8	28.3	16.1

Source: Presentation of Alexander Ozhegov at the FOA symposium on the Russian Defence Industry, Stockholm 20–22 Oct. 1993.

tion of defence equipment.[68] The probability is that most associations with more than 10 000 employees would have a much smaller number directly engaged in military production.

This heavy concentration in industry reflected the widespread lack of faith in the system of distribution and the risk of local shortages of materials and goods in spite of the preferential treatment the defence industry enjoyed across Central and Eastern Europe. Jacques Sapir has described this process of reducing reliance on supplies from distant locations as 'the production equivalent to hoarding stocks'.[69]

Restructuring of enterprise organization

In Central and Eastern Europe some of the most important defence manufacturers responsible for the assembly of larger military systems as well as their suppliers of components are deliberately trying to change their structure. It is important not to exaggerate the extent to which the restructuring process has been implemented in some Central European countries. Least progress seems to have been made in Russia and Ukraine although there are some regional 'islands' such as St Petersburg where more changes have occurred and which have been regarded as a form of laboratory for industrial restructuring.[70]

One approach is the 'atomization' of industrial units, a process by which large multi-product enterprises are being broken up into

[68] See, for example, note 19, the appendices.
[69] Sapir, J., *The Russian Defence Related Industries Conversion Process*, Centre d'Études des Modes d'Industrialisation, Paris, Oct. 1993, p. 12 (mimeographed).
[70] For example, Fedorov, B., 'Privatisation with foreign capital' eds A. Åslund and R. Layard, *Changing the Economic System in Russia* (Pinter: London, 1993).

smaller, more product-specific units.[71] A second approach is the consolidation of vertically integrated industrial units which control a significant proportion of the R&D and production cycle for any given system. This means combining the assets of what are at present independent units.

The breakup of large industrial units seems primarily confined to those enterprises which would prefer to leave the defence sector. This development has been hindered by the growth of intra-enterprise debt, which has had the effect of solidifying traditional ties with suppliers and making it more difficult for defence industry enterprises to form relationships with new suppliers on a commercial basis.

Examples of concentration through acquisition can be found across the region. There are cases of small component suppliers being absorbed by larger industrial units in the Czech Republic and Poland, as well as in Russia and Ukraine. While neither the Czech Republic nor Poland is likely to develop a major national defence industry, one or two medium-size companies may emerge in these countries. This may happen in order to exploit comparative advantages in a particular market niche—such as light utility helicopters in Poland or jet trainer aircraft in the Czech Republic. An additional motivation for this process has been to involve local suppliers to provide key components previously provided by suppliers with whom contacts have been broken.

In the Czech Republic the Aero holding company was established in 1991 by setting up 11 subsidiaries as joint stock companies. While the state retained a majority shareholding, about 35 per cent of shares are now owned by individuals or institutions.

Military aircraft production is concentrated in one of these subsidiaries, Aero Vodochody. However, the medium-term strategy of this subsidiary is also based on diversifying its activities. Future business will focus on: (a) military jet trainers and light attack aircraft; (b) design and development of small utility passenger aircraft; and (c) subcontracting work in civil aviation.

A similar pattern can be observed in the Slovak enterprise ZTS. Management responsibilities are currently being transferred from the headquarters down to individual manufacturing and service divi-

---

[71] For further discussion see Anthony, I., Claesson, P., Courades Allebeck, A., Sköns, E. and Wezeman, S. T., 'Arms production and trade', SIPRI, *SIPRI Yearbook 1994* (Oxford University Press: Oxford, 1994, forthcoming).

sions.[72] By the end of 1992, some of the most influential Slovak defence industrialists had reached the conclusion that under prevailing domestic and international conditions the defence industry could not be preserved.[73]

The ultimate intention of the ZTS enterprises is to create a diversified group in which overall revenue from civil production is increased (in 1991 around 51 per cent of revenues came from military sales). Eight divisions are being prepared to operate within the company. They include those devoted to construction equipment and military equipment; diesel engines and gearboxes; machine tools; special purpose machines; steel casting and forging; and forestry equipment. Another division—tractor engines—will be spun off as an independent company in which ZTS will remain the majority shareholder. Non-manufacturing service activities will also operate as separate divisions.

The primary motivation in this is to enable ZTS can establish itself as a financially viable privately owned company. Other objectives—maintaining employment and sustaining military production—are seen as secondary.

Another important change suggested by this likely reorganization concerns the traditional relationship between large associations responsible for the production of complex systems and their suppliers.

Aero has had difficulty maintaining its supplier networks. Some have been paid late and thus forced to go out of business. Many smaller industrial units which act as suppliers of components are now becoming increasingly dependent on Aero—some to the extent that Aero is their only customer. The same is true for ZTS which is unable to continue building the T-72 tank because of an inability to pay its suppliers.[74]

In Russia a new initiative taken by the MOD is to create 'financial industrial groups' which would combine many elements which are found in major corporations elsewhere in the world. These groups would bring together in a joint venture Russian units with design and production capabilities. The group would also include at least one

[72] Comments of Peter Magvasi, Financial Director, ZTS Martin, and Juraj Kovacik, Manager, Engine Division, ZTS Martin, at the SIPRI workshop on The Future of the Defence Industries of Central and Eastern Europe, 29–30 Apr. 1993.

[73] Magvasi, P. quoted in Kiss, J., 'Lost illusions? Defence industry conversion in Czechoslovakia 1989–92', *Europe–Asia Studies*, vol. 45, no. 6 (1993), pp. 1045–69.

[74] Fisher, S., 'The Slovak arms industry', *RFE/RL Research Report*, vol. 2, no. 38 (24 Sep. 1993), p. 38.

financial unit—such as a bank or an investment fund—as well as a trading organization to handle offset and counter-trade transactions. Finally, the preferred structure for such a group would include both military and non-military industrial units, with non-military production accounting for 50 per cent of sales or more. One of the first of these financial industrial groups was formed in 1993 bringing together the Almaz science and production association; the Fakel machine building design bureau; serial production plants in Moscow, Nizhniy Novgorod, Novosibirsk and St Petersburg; the Spetsvneshtekhnika State Foreign Economic Commission; the Oboronexport trading association; the Inkombank; and the Central Industrial Investment Check Fund.[75]

The process of industrial concentration is also apparently under way in Ukraine, at least in the aircraft industry. New aircraft construction is authorized by the Ukraine Aeronautical Industry Development Programme. This programme is actively seeking to involve Ukrainian manufacturers. Antonov, with overall responsibility for the programme, has discovered more than 150 new industrial partners in Ukraine many of which never previously participated in aircraft engineering. This is part of the effort to create a national aeronautics industry.[76]

Nevertheless, there is no intention in Ukraine of breaking its ties with Russia or other former collaborative partners. Given the integrated structure of defence production in the former WTO and the changing nature of the global defence industry, the future for most if not all enterprises is closely linked to the issue of industrial internationalization.

[75] Rudenko, V., 'Interview with Academician Boris Bunkin, '"Almaz" Science and Production Association General Designer and President of the "ROS" Company', *Krasnaya Zvezda*, 4 Sep. 1993, p. 4, in Foreign Broadcast Information Service, *Daily Report–Central Eurasia (FBIS-SOV)*, FBIS-SOV-93-173, 9 Sep. 1993, pp. 43–45.

[76] Comments of Oleg Bogdanov, Chief Designer, Antonov Design Bureau, at the SIPRI workshop on The Future of the Defence Industries of Central and Eastern Europe, 29–30 Apr. 1993.

5. International dimensions of industrial restructuring

Ian Anthony

I. Introduction

A number of analysts have drawn attention to a new phenomenon in the defence industry, namely the growing tendency to collaborate across national borders at all levels of the arms sales process: research, development, manufacturing and marketing. International collaboration has now extended to after-sales services such as service life extension programmes and mid-life upgrades of equipment.

For companies in Western Europe and North America internationalization is necessary for access to new markets, capital or technology. At a minimum it is becoming advantageous to team up with a local partner with official contacts and who understands national procurement regulations. These marketing functions are much better developed than collaborative production and the arms industry is a long way from resembling the form of industrial organization favoured in the automobile industry. However, the long-term trend in several of the major arms-producing countries of Europe and North America may be in the direction of retaining research, design and system integration at home while devolving assembly and component manufacture to regional partners. These regional partners may subsequently act as local maintenance and repair centres. The extent of this process—which has a 20-year history in civilian industrial sectors—should not be exaggerated. For the most part defence industries are still national in their orientation. Nevertheless, over the longer term this trend may lead to a shift in the structure of the global arms industry with implications for defence planning, technology development, arms procurement and efforts to prevent the misuse of military technology.

NATO collaboration was historically linked to the development of specific complex and technologically unproven weapon platforms which single countries could not develop alone. These projects were initiated by governments and production would take place in a designated company in each participating country. Recently companies have begun to collaborate across borders in a general sense rather than pursuing limited co-operation on one given system. Their collaboration is driven more by their own commercial agenda in an environment of shrinking military expenditure than by deliberate government policy. Indeed, some observers believe that whereas in the past government intervention stimulated cross-border company activity, governments now represent the main barrier to the development of international ties. Commercial pressures for industrial concentration are such that 'more liberal government policies would possibly result in a wave of international takeovers in the arms industry'.[1]

The dissolution of the Soviet Union and the division of Czechoslovakia have created international linkages from what were previously national industries. In these circumstances how and to what extent will countries of Central and Eastern Europe find their defence industries affected by the trend towards internationalization?

The defence industries of Central and Eastern Europe have been engaged in industrial collaboration across borders—albeit of a very different type from that described above. Within the WTO a division of labour was organized by the Soviet Union as a means of supplementing its own military industrial capacity.[2] Older systems no longer in production in the Soviet Union but still used by the armed forces were serviced by spare parts produced in non-Soviet WTO countries.[3] The Soviet Union to a large extent determined the types of equipment produced by and the level of technical competence of Central European industries.

Non-Soviet production of missiles in the WTO was confined to Romania where short-range air-to-air missiles of Soviet design were produced. Similarly, production of electronic systems was limited outside the Soviet Union. The capability for the development of electronic and telecommunication systems in Central Europe was limited and largely concentrated in Bulgaria, Hungary and the German Democratic Republic. For the most part production in the non-Soviet

[1] Sköns, E. and Wulf, H., 'The internationalization of the arms industry', Paper prepared for the workshop on The Arms Trade and Arms Control in the Post-Cold War World, American Academy of Political and Social Science, New York, 4–5 Nov. 1993.

[2] Kiss, J., 'Military production and arms trade in Hungary', Paper prepared for the SIPRI workshop on The Future of the Defence Industries of Central and Eastern Europe, 29–30 Apr. 1993.

[3] *Warsaw Voice*, 8 Dec. 1991, pp. B4–5.

WTO countries was heavy engineering, producing relatively un-sophisticated systems developed in the 1970s or earlier.

The dissolution of the WTO had a dramatic impact on defence industrial co-operation. Intra-WTO sales accounted for a significant element of overall defence industrial activity across Central and Eastern Europe. The remainder of this chapter is confined to the discussion of those forms of internationalization other than cross-border sales of finished goods which are relevant to the Central and East European situation.

There seem to be three elements to internationalization in the central and East European context. First, enterprises of Central and Eastern Europe wish to link with foreign partners as part of their strategy for survival through diversification. This requires foreign direct investment but seems unlikely on a large scale for reasons discussed below.

Second, some local producers are likely to exploit their comparative advantage in doing business with their own ministry of defence. As noted in chapter 3, Central and East European countries are unlikely to allocate significant resources for procurement in the immediate future. However, these countries would eventually like to re-build and modernize their armed forces. Several face the challenge of a complete re-orientation in defence policy to reflect new state boundaries. If and when their economies recover sufficiently to permit an increase in military expenditure these countries may become a significant market for defence manufacturers. Teaming with a local partner may be a condition for doing business in Central and Eastern Europe.

In the interim there are efforts to introduce new technologies into major weapon platforms through the purchase of foreign sub-systems and components. The modified platforms are initially more likely to be offered for export than to the government of the manufacturer.

Third, efforts have been made to reconstruct industrial relationships ended by the breakdown of intra-WTO co-operation.

II. Internationalization as a form of diversification

In recent years there have been a significant number of start-ups in industrial joint ventures between countries of the Organisation for Economic Co-operation and Development (OECD) and countries of Central and Eastern Europe (what used to be called East–West joint ventures). In 1987—the year after the Soviet Union enacted legislation permitting joint ventures—23 collaborative ventures were registered. By April 1991 there were 3400 registered joint ventures on the territory of the former Soviet Union of which 1188 had started operations.[4]

For many of the defence industries in Central and Eastern Europe internationalization is seen as a strategy for reducing dependence on sales to military customers by using foreign capital and technology to transform production capacities. However, identified joint ventures are heavily concentrated in eight non-military sectors which together accounted for around 65 per cent of the total in 1991. These sectors are computers, consulting, consumer items, chemicals, machinery, food processing, tourism and hotel construction.[5]

Joint ventures involving participation by enterprises from the defence industry are not individually isolated in the available data but comparatively few seem to have been established. By late 1992 there were only 180 joint ventures involving enterprises with defence-related production in Russia and only 220 for the whole CIS.[6] Of these joint ventures US, German, British and Italian companies accounted for around 40 per cent.

Links between civil manufacturers in Central and Eastern Europe and extra-regional partners have a long history. Such partnerships were never excluded by the COCOM embargo and, where there were mutual economic benefits, co-operation occurred throughout the cold war.[7] Nevertheless, many obstacles to foreign joint ventures remain.

Joint venture activity in the defence sector

Most discussions with the defence-related sector seem to have focused on the space and aircraft industries. For example, table 5.1

[4] United Nations Centre on Transnational Corporations, *Accounting for East-West Joint Ventures* (United Nations: New York, 1992).

[5] O'Boyle, T. F., 'Soviet break-up stymies foreign firms', *Wall Street Journal*, 23 Jan. 1992.

[6] Cooper, J., *The Conversion of the Former Soviet Defence Industry* (Royal Institute of International Affairs: London, 1993), p. 35.

[7] Alex McLoughlin, Head of Trade Affairs for the company ICL, from his presentation to the NATO–CEE Conversion Seminar, Brussels, 20–22 May 1992. ICL has been doing business in the region for over 30 years. The speed with which East–West trade resumed after 1945 was one of the origins of the embargo. See Adler-Karlsson, G., *Western Economic Warfare 1947–67: A Case Study in Foreign Economic Policy* (Almqvist & Wiksell: Stockholm, 1968).

illustrates that all of the large Russian aircraft design bureaus—including Ilyushin, Kamov, Mikoyan, MIL, Sukhoi, Tupolev and Yakovlev—have had discussions with foreign partners as have the jet engine manufacturers Klimov, Lyulka and Perm.

The agreements outlined in table 5.1 indicate that most projects are not aimed at military production but at passenger or freight transport in the CIS countries. Oleg Bogdanov of Antonov has noted:

In the period 1982–85 an abrupt increase of cargo traffic volume took place, with up to 6 per cent growth yearly. Growth decelerated in 1986–88, since when cargo traffic volume started falling. The situation in cargo traffic accurately reflects the condition of industry. We forecast stabilization in 1994, and subsequent cargo traffic increases. . . . New cargo routes are being investigated and will be implemented in the nearest future. Having examined those trends we think that cargo traffic rates will not just stay on the level of the 1980s, but will grow further.[8]

Foreign companies are seeking to gain a foothold in the civil aircraft market with (rather than at the expense of) local partners because the market is closed to foreign companies unless they have government authorization. As one US executive has noted: 'any way you look at Aeroflot, you look at it and salivate, but no one knows how to lock up a deal with them, especially in a world market that is already oversized'.[9] In other areas (and perhaps these hold the greatest promise from a commercial perspective) there are some attempts to introduce new products which are related to the traditional expertise of the Russian partner—for example, the use of aircraft engines for generating electricity.

In terms of strategic positioning many foreign companies would be interested in partners with marketing and distribution skills whereas military producers have been used to dealing with a single customer—the government. Therefore the civil aircraft sector is probably a special case in that it duplicates elements of the military market. It is a state monopoly involving the sale of a small number of large and complex systems each with a very high unit development cost and unit purchase price. These characteristics are present in relatively few civil markets.

[8] Oleg Bogdanov, Chief Designer, Antonov Design Bureau, at the SIPRI workshop on The Future of the Defence Industries of Central and Eastern Europe, 29–30 Apr. 1993.

[9] Tom Culligan, Vice President for Programmes and Marketing, McDonnell Douglas, interviewed in *Russian Aerospace and Technology*, 13 July 1992, p. 6.

Table 5.1. Selected joint ventures with Russian aircraft industry, 1991–93

Russian partner	Foreign partner	Country	Date[a]	Comment
Aviapribor	Sextant Avionique	France	1992	Joint development of avionics for civil airliners
Aviastar	Hunting	UK	1991	To produce interiors for TU-204-100/200 airliner
Buran	Alenia Thomson CSF Westinghouse	Italy France USA	1993	Development of air traffic control radars
Ilyushin	Pratt & Whitney MTU Collins	Canada FRG USA	1990	Re-engine and new avionics for IL-96M airliner
Ilyushin	CFM International	France/ USA	(1993)	Re-engine of Il-86 airliner
Ilyushin	Allison	USA	1993	Re-engine of Il-114 airliner
Ilyushin	Collins	USA	1993	Development of cockpit for Il-96M airliner
Kamov	Rolls Royce	UK	1990	Re-engine of Soviet Ka-62 helicopters
Kamov	Eurocopter/ SNECMA	France	(1992)	Joint development of light helicopter
Kamov	Group Vector	Switz.	(1992)	Plan to build Ka-50 in Greece for export
Kamov	Allison	USA	1992	Engine for Ka-126 helicopter
Kamov	Daewoo	South Korea	(1993)	To develop a helicopter for crop spraying
Klimov MiL Kazan	Eurocopter	France	(1993)	Eurocopter to fit cockpit, avionics, passenger systems and market civilian Mi-38
Klimov Rusjet	AeroSud Pratt & Whitney	S. Africa Canada	1992	Development of a light utility turboprop aircraft
Klimov	Pratt & Whitney	Canada	1993	To develop engines for Ilyushin air freighters
Lyulka	Rolls Royce	UK	1990	To develop an engine for business jets
MiG	Promavia Garrett Bendix King Rolls-Royce	Belgium USA UK	1992	Collaboration on the ATTA 3000 jet trainer
MiG	Rediffusion	UK	1992	To develop simulators for MiG-29 and MiG-31
MiG	Daewoo	South Korea	1992	To produce aircraft braking systems
MiG	Dassault	France	1993	To produce assemblies for Falcon 50/900 business jet
MiG	Messier Bugatti	France	1993	To develop hydraulic pumps for MiG-AT

Russian partner	Foreign partner	Country	Date[a]	Comment
MiG	SNECMA Turbomeca	France	1993	To develop MiG-AT jet trainer
	MTU	FRG		
MiG	Thomson CSF	France	1993	To develop a MiG-21 upgrade
MIL	Brooke Group	USA	1992	To develop global spare parts support for MIL helicopters
MIL	Daewoo	South Korea	(1993)	To market civil versions of MIL helicopters in Asia
National Institute of Airborne Avionics	Allied Signal	USA	1992	To develop cockpit for Be-200 and Yak-242 civil aircraft
Penza	CAE Electronics	Canada	1993	To develop simulators for several Russian aircraft types
Perm Motors	Pratt & Whitney	Canada	(1992)	To produce upgraded turbofan engine for TU-204 airliners
Perm Motors	SNECMA/ General Electric	France/ USA	1992	To produce upgraded turbofan engine for TU-204 airliners
Sukhoi	Grumman	USA	1990	To develop a business jet
Sukhoi	GEC	UK	(1992)	GEC to fit avionics to SU-27 fighters for export
TEI	BDM International	USA	1993	Marketing company for Russian products in the US
TsAGI	Boeing	USA	1992	Boeing to use TsAGI test facilities
TsAGI	Mitsubishi	Japan	1993	Mitsubishi to use TsAGI test facilities
Tupolev	Rolls Royce	UK	1991	TU-204-100/200 unveiled at 1993 Paris Air Show
Tupolev	Aerospatiale	France	1992	To develop TU-334 airliner
Tupolev	BMW/ Rolls Royce	Germany UK	1992	Re-engine of TU-334 airliner
Yakovlev	Aerospatiale Alenia	France Italy	(1993)	Yakovlev to licence produce ATR-42/72 airliner
Yakovlev	Israel Aircraft Industries	Israel	1993	To develop Astra 4 business jet
Yakovlev	Textron Lycoming	USA	1992	To develop a new airliner
Yakovlev	Rockwell Collins	USA	1992	To develop a jet trainer
Yakovlev	Allied Signal	USA	1993	Avionics for Yak-142 airliner
Yakovlev	Aermacchi	Italy	1993	To develop Yak-130 jet trainer

[a] ( ) means project not finalized at the time of writing.

Source: SIPRI archives.

Joint production require a major injection of capital in most cases. As a US executive observed of Russian manufacturers: 'they don't really bring much to the table. . . . We've sent several delegations of executives over there and have toured their factories. They just aren't able to pay for anything.'[10] Major direct investment to develop manufacturing facilities 'from the ground up' seems unlikely. While labour costs for skilled workers are currently low by West European and North American standards the value of this advantage is disputed. While some industrialists believe it will last at least until the end of the 1990s, others point out that wage costs can rise quickly and there has to be a more durable basis for collaboration to make an investor commit large sums to a project.[11]

Many joint projects in Russia have had political rather than commercial roots. For example, Sir Fred Catherwood, Vice Chairman of the European Parliament Committee on Foreign Affairs and Security, gave the desire for a 'stable, friendly and democratic Eastern Europe' and a 'mutual and simultaneous reduction' in levels of military expenditure across Europe as the main justifications for financial aid.[12]

Projects have been designed in accordance with the security interests of the foreign government providing financial support—assistance with environmental cleanup, weapon destruction, nuclear plant safety, and constructing housing for troops withdrawn from former WTO countries. These programmes, while worthwhile on their own terms, can play little role in developing self-sustaining economic partners in Central and Eastern Europe.

Financial support by foreign governments for civil projects with a straightforward commercial rationale is more difficult to justify and must comply with existing multilateral rules on trade finance and export subsidy. Consequently, if there is to be diversification through internationalization in the defence industry, it is likely to result from industry-led initiatives. However, for the reasons noted above, defence industries rarely make attractive partners for foreign

---

[10] See Culligan (note 9), p. 6.

[11] Comments by Adam Stranák, Aero Vodochody and Oleg Gapanovich, St Petersburg City Council, Military Industry and Conversion Commission, at the SIPRI workshop on The Future of the Defence Industries of Central and Eastern Europe, 29–30 Apr. 1993. One of the first effects of a successful joint venture is an increase in local salaries.

[12] *Aid Instead of Arms—A Practical Proposal of East–West Military Conversion*, Joint Hearings before the Committee on Foreign Affairs and Security; Committee on Budgets; Committee on Economic and Monetary Affairs and Industrial Policy; Committee on External Economic Relations, European Parliament, 21 Apr. 1993.

investors. There may be individual successes but diversification strategies based on international partnerships seem unlikely to play a major role in the restructuring of the defence industries of Central and Eastern Europe.

III. Teaming with foreign suppliers to enhance military capabilities

There is continued demand for military equipment among the countries of Central and Eastern Europe. Central European countries found that their capacity for national defence was degraded when departing Soviet/Russian forces took with them equipment, expertise and manpower. In order to fill gaps in equipment and force structures Central and East European countries have examined several alternatives. One is industrial teaming with foreign producers to develop new systems. A second is the direct import of foreign systems. A third option that has been examined is the re-nationalization of defence production. A fourth is the modification of existing systems with foreign assistance.

The first option—industrial teaming aimed at joint development of new systems—has been prevented by a combination of lack of local financing and global over-capacity in the defence industry. For example, the French company SNECMA and the Russian enterprise Klimov explored the joint development of a fighter aircraft engine but the project failed when Klimov was unable to contribute any resources. A project established in 1993 to develop a major platform would be unlikely to have a product ready for marketing before the year 2000. In the intervening period the project would drain scarce resources from the Western partner with no guarantee that any market would exist for the product at the end of the development phase.

Some products may find a market niche. For example, Rediffusion—a British subsidiary of US company Hughes Electronics—has an agreement with MiG to develop a training simulator for two advanced Russian fighter aircraft—the MiG-29 and the MiG-31. Such a simulator may have a market given that the MiG-29 is operated by 13 countries (excluding the air forces of the CIS where more than 600 aircraft are in service).

Direct imports of major military systems are politically unacceptable to Russia regardless of economic considerations because of the desire to retain defence industrial activity under national control.

Indeed, there is some evidence of import substitution to replace critical elements of the defence industrial base located in parts of the former Soviet Union which are now independent.[13] However, the possibility of reducing dependence on Soviet equipment was examined by Central European countries in 1990–91 after the members of COCOM began to modify the embargo.

Polish military pilots evaluated the US F-16 fighter and the French Mirage-2000 and discussed the terms of purchasing the JAS-39 Gripen fighter with Sweden.[14] Hungary discussed the JAS-39 fighter with the Swedish authorities as well as investigating the installation of air defence systems to provide all-around coverage of Hungarian air space.[15] The Defence Minister of the Czech Republic suggested in April 1993 that Central European countries should have examined joint procurement of systems such as air defence radars or telecommunications systems both to ensure compatibility and in order to reduce costs.[16] No such discussions occurred and there is no current prospect of direct sales of major systems to Central European countries. The only significant transfer of military items to a former WTO country—the sale of 118 Identification Friend or Foe (IFF) systems to the Hungarian Air Force in late 1992—occurred under unique circumstances. NATO airborne early-warning aircraft have been operating from Hungary monitoring the airspace of the former Yugoslavia. Air forces in the former Yugoslavia are operating the same aircraft types as Hungary—notably the MiG-21—and it is necessary for NATO aircraft to be able to distinguish between the two.[17]

Some analysts have observed that a re-nationalization of the defence industry is currently taking place.[18] In Poland the local producer of the T-72 tank, Bumar Labedy, together with the Military Institute of Armament Technology, has developed Polish versions of systems previously obtained from the Soviet Union, including

[13] Remarks of Julian Cooper at the Försvarets forskningsanstalt (Swedish National Defence Research Establishment) (FOA) seminar on the Future of Russian Defence Industry, Stockholm, 21–22 Oct. 1993.

[14] *Defense News*, 18 Feb. 1991, p. 17; *AAS-NL Milavnews*, Oct. 1991, p. 20.

[15] *Defense News*, 14 Oct. 1991, p. 1; *Flight International*, 23–29 Oct. 1991, p. 4; *Defense News*, 22–28 June 1992, p. 42.

[16] Antonin Baudys, quoted in Hitchens, T., 'Defense Co-operation confounds central Europe', *Defense News*, vol. 8, no. 16 (26 Apr.–2 May 1993), p. 1 and 29.

[17] *Arms Transfer News*, vol. 93, no. 1 (1993).

[18] Maciej Perczynski, Polish Institute for International Affairs, at the SIPRI workshop on The Future of the Defence Industries of Central and Eastern Europe, 29–30 Apr. 1993.

engines, armour, thermal sights and a radar warning receiver. The tank incorporating these Polish sub-systems is designated the PT-91 Hard.[19] However, the decision to begin development of Polish electronic systems was taken only after the United States refused Israeli company Elop permission to assist Poland in the programme.[20] Similarly, a Polish turbofan engine has been developed for the I-22 Iryda jet trainer made by the PZL enterprise at Rzeszow.[21] In each case the local solution was a product of necessity rather than choice. The Polish Government evaluated foreign equipment to replace sub-systems previously obtained from the Soviet Union. However, even limited upgrading with West European equipment moved equipment out of the price range the Ministry of Defence could afford.

Upgrades and modifications

Of the forms of industrial collaboration in the military field the modification of Central and East European platforms with foreign sub-systems seems the most likely to occur. In an environment where new programmes are scarce, upgrading and modifying existing systems are natural options for arms producers outside Central and Eastern Europe seeking to support their design and manufacturing capacities against a background of shrinking military expenditure.

Several countries of Central and Eastern Europe have explored this kind of project as a means of modernizing their own armed forces. However, they have been deterred by the cost of sub-systems. Electronic systems may amount to 30–40 per cent of the cost of a new platform while the power unit (engine and transmission) can be equally expensive. However, the place of the Soviet Union as a major arms exporter during the 1970s and 1980s has meant that many countries of the world have armed forces built around Soviet equipment. This dependence will continue for many years, especially given the depressed level of military expenditure in most parts of the world.

Examples of bilateral co-operation are now beginning to appear. In Russia the Sukhoi design bureau and British company GEC have discussed collaborating on a version of the Su-27 with a redesigned cockpit to include a new head-up display, multi-function radar and

[19] *Warsaw Voice*, 25 Apr. 1993, pp. 12–13; *Jane's Defence Weekly*, 17 Oct. 1992, p. 11; *International Defense Review*, June 1993, p. 489.

[20] *Intelligence Newsletter*, no. 223 (2 Sep. 1993), p. 5.

[21] *Interavia Aerospace World*, May 1993 p. 11.

fly-by-wire controls. MiG has had discussions with French company Thomson CSF with a view to producing a modernized MiG-21 fighter aircraft. The upgrade would include new radars, electronic warfare systems and a redesigned cockpit incorporating new displays and navigation systems.[22] This aircraft—of which around 6000 were built by the Soviet Union—is in service in 38 countries, many of which will keep it as the mainstay of their air forces for at least the next 10 years. Such an arrangement has the attraction that many potential customers are already familiar with MiG-21 repair and maintenance and are able to carry out much of the work.[23]

Enterprises in Central and East European countries can contribute expertise to a joint venture with a West European or North American partner and need not contribute financially. However, as noted above, systems developed in this way may still be beyond the means of Central and East European governments. Third-party involvement is being sought to finance collaborative development. The MiG/ Thomson modernization package was offered to India during a visit by a delegation of the French aerospace industries association Groupement des Industries Françaises Aéronautiques et Spatiales (GIFAS) in April 1993.[24]

An explicit element of licence agreements within the WTO was that no product would contain Western imported parts.[25] Whether these agreements are still legally valid is not clear but in the new political environment they are no longer being applied. The potential for modernization of Soviet systems has been widely noted in the arms industry and a version of the MiG-21 named the 'MiG-21-2000' is being developed by Israeli companies Israel Aircraft Industries (IAI) and Elbit. This aircraft has been selected by Romania as the basis for its air force modernization.[26] Aero Vodochody of the Czech Republic has held discussions with Allied Signal of the United States about developing a MiG-21 upgrade package for Egypt which could then be used by the Czech Air Force.

There have been fewer agreements in the area of land systems than there have been for combat aircraft. However, in 1992 and 1993 French companies have established ties with enterprises engaged in

[22] Aviation Week & Space Technology, 5 July 1993, p. 17.
[23] International Defense Review, June 1993, pp. 445–50.
[24] Asian Recorder, 16–22 Apr. 1993; Defense News, 10–16 May 1993, p. 16.
[25] See Kiss (note 2).
[26] Interavia Air Letter, 25 May 1993, p. 1.

the manufacture of the T-72 tank in the Czech Republic, Poland and Slovakia with a view to producing an upgrade package.[27] The contacts have been co-ordinated by the French government agency SOFMA (the Société Française de Matériels d'Armament). This upgrade includes the addition of new fire control systems and engines and could be offered not only to governments in Central Europe but also to the operators of the T-72 elsewhere in the world. As of late 1992 20 000–25 000 T-72s had been produced and the tank is operated by more than 25 countries.[28]

IV. Restoring defence industrial ties between Central and East European countries

Although Soviet political direction of international collaboration between the defence industries of Central and Eastern Europe has now disappeared, the pervasive presence of Russian technology throughout the region represents a significant legacy from the past. These governments now face a severe problem of providing their armed forces with the necessary maintenance, repair and logistical support for many items retained in existing inventories.

Central European countries find themselves with very similar types of equipment in the inventories of their armed forces. Economic constraints prevent the purchase of new equipment and, in the case of some specific equipment types, the overall number of systems permitted is restricted by the 1990 CFE Treaty. In these circumstances industry is unlikely to be receiving orders for the construction of new equipment. Nevertheless, maintaining the equipment already in service through collaborative arrangements between countries may serve a useful purpose for both the armed forces and the defence industry.

In spite of this interdependence and the apparent scope for industrial co-operation, there is no political framework for restoring regional or sub-regional collaboration between defence industries in Central and Eastern Europe. Visegrad officials met in Cracow, Poland on 6–7 September 1993 to discuss supplying each other with military equipment and spare parts, ways of increasing the volume of orders

---

[27] *Defense News*, 20–26 Sep. 1993, p. 1. SOFMA is the agency responsible for facilitating arms exports through the arrangement of financing and credit.

[28] The low estimate is contained in *World Military Vehicles Forecast* (Forecast International: Newtown, Conn., 1993); the higher one in *Jane's Armour and Artillery 1992–93* (Jane's Information Group: Coulsdon, UK, 1992), p. 24.

from each other and collaboration to make technological improvements in their weapons.[29] However, the Visegrad Group, for example, has not emphasized defence industrial ties. On the contrary, at the political level it is aimed at achieving the earliest possible entry into both the European Union and NATO.

While the Visegrad Group is one possible organizing body for government-to-government discussions, if the primary intention is industrial co-operation there is no reason to exclude countries like Ukraine or Belarus from the process. Table 5.2 indicates that there are some efforts to restore bilateral co-operation between governments. Projects with non-Russian former Soviet republics could form the basis for bilateral industrial collaboration.[30]

By the time Central European economies begin to grow at a rate that allows new equipment to be bought some industrial capacities will have been lost. How remaining capacity will be distributed between manufacturing industry and the armed forces is not clear. In the Czech Republic defence producers are trying to persuade the government to put maintenance and support work carried out by the armed forces out to tender.[31]

The lack of attention paid to collaboration within the region at the political level is a cause of frustration to many industrialists who see the destruction of the WTO internal market as a mistake which created far more problems than it solved. Government-to-government contacts and agreements were the central element in the functional distribution of labour between countries with planned economies. These official contacts were scaled back or cancelled as Central European countries asserted their political independence with severe consequences for industry. Moreover, the policy of developing closer integration with NATO and the European Community has moved forward slowly and uncertainties remain about the nature of any future relationship with these organizations and the timetable for achieving changes.

---

[29] Radio Free Europe/Radio Liberty, *RFE/RL News Briefs*, vol. 2, no. 37 (6–10 Sep. 1993), pp. 13–14.

[30] Peter Magvasi, Financial Director, ZTS Martin, at the SIPRI workshop on The Future of the Defence Industries of Central and Eastern Europe, 29–30 Apr. 1993.

[31] This approach has been adopted in the UK (although not without controversy) and a similar arrangement is now under discussion in the USA.

Table 5.2. Government-to-government framework agreements for defence industrial co-operation, 1991–93[a]

Country	Date of agreement	Source
Czech R.–Hungary[b]	(see note)	*East Europe Intelligence Report*, 25 Mar. 1993
Czech R.–Poland	Feb. 1991	*RFE/RL Research Report*, vol. 2 no. 14, 2 Apr. 1993, p. 31
Hungary–Poland	Feb. 1991	*RFE/RL Research Report*, vol. 2 no. 14, 2 Apr. 1993, p. 31
Hungary–Russia[c]	Sep. 1993	*RFE/RL News Briefs*, vol. 2, no. 40, 27 Sep.–1 Oct. 1993, p. 16
Hungary–Slovakia[d]	Apr. 1993	*RFE/RL News Briefs*, 26–30 Apr. 1993, p. 10
Poland–Russia	July 1993	*RFE/RL News Briefs*, 5–9 July 1993, p. 17
Poland–Slovakia	Feb. 1991	*RFE/RL Research Report*, vol. 2 no. 14, 2 Apr. 1993, p. 31
Poland–Ukraine	Feb. 1993	*RFE/RL Research Report*, vol. 2 no. 14, 2 Apr. 1993, p. 31
Russia–Slovakia	Aug. 1993	*RFE/RL Research Report*, vol. 2 no 38, 24 Sep. 1993, p. 38
Russia–Ukraine	Jan. 1993	*RFE/RL Research Report*, vol. 2 no 25, 18 June 1993, p. 40

[a] In each case military-technical co-operation is mentioned as an element in the agreement but the precise content of the agreement is not known.

[b] In Mar. 1993 the defence ministers of the Czech Republic and Hungary announced that there were unable to expand co-operation in the development and production of military equipment.

[c] On 29 Sep. 1993 Hungarian Defence Minister Lajos Fur discussed co-operation in Moscow.

[d] In Apr. 1993 Hungary submitted a draft agreement on co-operation in the development and production of military equipment to Slovakia during the visit of the defence minister to Bratislava.

In these circumstances the financial and technical advantages of collaboration within Central and Eastern Europe may outweigh—or at least match—the urge to join other organizations. A Hungarian government official has observed: 'joining NATO is an objective for the future and realization is conditional on resolving the tensions existing in the region and recovering from economic difficulties—including those of their defence industries. Therefore, Hungary attributes great significance to the co-operation agreed on at Visegrad, which may represent a viable road to recovery through reliance on local resources'.[32]

The motive for international co-operation from the government perspective is the need to achieve cost-efficient procurement and to rationalize the maintenance and repair of defence equipment in an effort to compensate for the current low level of military expenditure. From an industrial perspective the motive is increasing the potential market beyond the national Ministry of Defence. The process underlines the fact that individual producers in Central Europe do not easily fall into clear categories of competitor or collaborator. Rather, since they face similar problems and share similar human, technical and financial resources, there are opportunities for both competition and co-operation. At the moment the competitive dimension of defence industrial behaviour is most obvious in the export market.

[32] Laszlo Kovacs, Director General, Ministry of International Economic Relations, Hungary, at the SIPRI workshop on The Future of the Defence Industries of Central and Eastern Europe, 29–30 Apr. 1993.

6. Arms exports

Ian Anthony

I. Introduction

From the late 1940s to 1990 the global arms trade was dominated by two superpowers which used arms transfers to advance their position *vis-à-vis* one another. It would be difficult to exaggerate the impact of the collapse of the former Soviet Union and the change in US–Soviet/Russian relations which stemmed from it for the international arms trade.

The end of the cold war removed the competitive ideological dimension from US–Russian relations and in these conditions neither country is now prepared to offer to grant military assistance to clients.[1] This has led to the weakening of bilateral relationships between countries of Central and Eastern Europe and recipients in developing countries which accounted for a major percentage of the total arms trade for most of the 1980s. Afghanistan, Cuba, Nicaragua, North Korea, Syria and Viet Nam have effectively been eliminated as markets by the ending of the programme of export subsidy. The difficulty of financing arms imports is likely to become the most significant restraining factor in the global arms trade.

In place of ideological competition, the USA and Russia have recently become partners in efforts at regional conflict resolution. Bilateral US–Russian or multilateral agreements now prohibit arms sales by either country to Afghanistan, Angola and Cambodia. In the United Nations the USA and Russia have co-operated to introduce mandatory embargoes which have closed several other important markets for weapons from the Central and East European arms industries, namely Iraq, Libya and the former Yugoslavia.[2]

The manner of the collapse of the former WTO was extremely disruptive for industries. In 1991 and 1992 there was a precipitous

[1] The only significant exception to this is the continued US military assistance to Egypt and Israel.
[2] The issue of the regulation of arms exports is discussed in chapter 4 as an element of the new approach to government regulation of industrial activity. Chapter 4 is confined to the role of Central and East European countries in the international arms market.

reduction in the scale of arms transfers and technological co-operation between the countries of Central and Eastern Europe. In some cases contracts were literally cancelled overnight and equipment which had already been produced was either never delivered or delivered and never paid for.

The end of the cold war has permitted some bilateral arms transfer relationships previously prohibited for ideological reasons to be opened or, in the case of Russia and China, reopened. The transfer of helicopters and armoured vehicles from Russia to Turkey is an example of a relationship made possible by the end of the cold war. As described in chapter 5, some forms of international industrial collaboration in the area of defence production have also become possible in the post-cold war period—notably co-operation between Israel and the Czech Republic, Poland and Romania.

For many years it was impossible to have access to any data measuring the volume (that is, the movement of goods) or the value (that is, the financial flows associated with the movement of goods) of the arms trade from official sources. The only government source of comparative data—the US State Department—provided estimates, except in the case of the USA itself. This is gradually changing and it is now possible to obtain official data of both types (although in neither case are available data comprehensive).

The section below considers the official data which are now available concerning the arms trade. As the section shows, the data are by no means straightforward or without problems, which are highlighted in the text.

II. Official data on the value of the arms trade

Exports from the former Soviet Union represented an enormous industrial production effort. Between 1945 and 1989 China received over 3000 Soviet military aircraft and helicopters of all types. Iraq received around 2000; Syria and Egypt around 1500 each; India 1400; Poland 1300; the German Democratic Republic and Afghanistan around 1200 each and Czechoslovakia around 1000. However, while the number of systems transferred has fallen, this does not necessarily indicate a reduction in financial flows. It may be that more income is generated by smaller volume of transfers because of changes in the term of trade. Andrey Kokoshin, Russian First Deputy Minister of Defence, has explained the reasoning behind this possibility.

It has to be acknowledged that the traditional arms markets, which we have now lost, were to a considerable extent not real markets. The country supplied weapons abroad often effectively free of charge and in vast quantities. That was the case with Ethiopia, for example, where whole division complexes were supplied and lost in just a few days. Russia is owed colossal amounts of money by Syria and Algeria. The problem of settlement with them has still not been resolved.[3]

A similar point has been made by the Czech aircraft company Aero Vodochody where a major reduction in sales between 1987 and 1992 does not imply a reduction in profitability. The chairman of Aero Vodochody has said 'with half of the previous production, profit levels will be six to nine times higher'.[4] Aero Vodochody has succeeded in finding customers such as Egypt and Thailand able to pay in hard currrency for its L-39/59 series of trainer aircraft which are acknowledged to be highly competitive in their product sector. However, this success is likely to be one of a few exceptions rather than typical for Central and East European arms suppliers.

Among the members of the WTO, arms transfers were considered as an element of foreign and defence policy and as such treated as a political rather than a trade issue. Nevertheless, the collapse of the CMEA contributed to a rapid decline in trade which has produced an economic crisis for industrial production of all kinds in Central and Eastern Europe.

Within the WTO inter-state financial transfers associated with arms transfers and defence industrial co-operation were agreed on the basis of convertible roubles using an exchange-rate fixed by negotiation between the various central planning agencies in the framework of the CMEA rather than through the operation of the market.[5]

Similar arrangements were made between the former Soviet Union and major arms recipients. The primary beneficiaries of credit arrangements were Algeria, Iraq, Libya, Syria and India—which was

[3] Gorokhov, N., 'Inteview with Andrey Kokoshin, Russian First Deputy Defense Minister', *Rossiyskiye Vesti*, 23 July 1993, p. 2, in Foreign Broadcast Information Service, *Daily Report–Central Eurasia (FBIS-SOV)*, FBIS-SOV-93-142, 27 July 1993, p. 34.

[4] *Interavia Aerospace Review*, June 1992, p. 41.

[5] A useful summary of CMEA trade relations is 'The Collapse of Trade Among Former Members of the CMEA', a survey prepared by staff of the International Monetary Fund contained in *World Economic Outlook* (IMF: Washington, DC, Oct. 1991).

able to make purchases both in exchange for commodities (the Soviet Union maintained a rupee account used to purchase goods in India) and against a credit account established at the Indian central bank. Credit was repayable over up to 17 years with an interest rate of around 2 per cent. However, since neither the rupee nor the rouble is a convertible currency the true balance of Indo-Soviet trade is also almost impossible to establish.[6]

In December 1990 the Soviet Union prohibited barter trade with former CMEA partners and, with the termination of the CMEA, all of the countries of Central and Eastern Europe decided to conduct foreign trade on the basis of hard currency payment. By mid-1991 the Soviet Union had lifted the prohibition on barter and by 1994 all Central and East European countries had retreated from their decision to trade only in hard currency, having found it almost impossible to conduct trade on this basis.

As noted in chapter 5, by 1993 elements of WTO defence industrial co-operation as well as arms transfer relationships had been re-established without accompanying hard currency payments. Arms transfers to China and India agreed in 1992 were also part of broader economic packages that included non-military industrial goods. Therefore, although most of the recent official estimates for the value of the arms trade have been given in US dollars, whether the values declared will correspond to money received cannot yet be established, even by the parties to the agreement.

The decisions by Hungary and Slovakia in 1993 to accept MiG-29 fighter aircraft and other military equipment from Russia as partial settlement of Russian debt also reflects an interesting new development in financing. The process through which the level of bilateral debt was established and that for establishing the value of the goods transferred both appear to have been arbitrary.[7]

---

[6] This became a major issue between India and Russia during President Yeltsin's visit to New Delhi in Jan. 1993. Russia, arguing that the oil, natural gas and capital goods transferred to India had been grossly undervalued, claimed a major debt was owed by India. India, meanwhile, pointed to the fact that in bilateral trade agreements the rouble was significantly overvalued. In the end a compromise was reached based on elements taken directly from past arrangements, including significant barter and long-term credit.

[7] Although in both cases the aircraft transferred were produced in the past two years—that is, after the beginning of the Gaidar economic reform programme—administrative price controls have not been lifted for the defence sector.

Russia

Russian data on the value of arms exports are beginning to become available but precisely how these data were compiled remains unclear. Under conditions where few if any new agreements involve hard currency payments it is very difficult to evaluate the financial flows associated with recent arms transfers.

In September 1992 G. Yanpolskiy, general director of the Defence Industry Department of the Ministry of Industry stated that while exports accounted for 30 per cent of sales by the Russian defence industry in 1991 they accounted for 7.2 per cent of sales in the first half of 1992.[8] In August 1993 Yanpolskiy stated that the value of arms exports from Russia in 1992 was $1.3 billion, reduced from a value of $6 billion recorded for 1990.[9]

Of the official data released the most detailed were provided by the Ministry for Foreign Economic Relations which released to the public data that had previously been given to the other permanent members of the UN Security Council in the context of the discussion of arms control in the Middle East. According to these data the value of arms deliveries by the former Soviet Union in 1991 was $1.55 billion of which $20 million was in the form of grants.[10] In November 1992 Peter Aven, then Russian Minister of Foreign Economic Relations, told the Russian Supreme Soviet that the value of Russian arms sales for 1991 was $7.8 billion—a reduction from a high point of $23 billion in 1989. According to Aven, the estimated value of sales for 1992 was $3 billion.[11]

It is possible that Aven was referring to the value of new agreements rather than deliveries of equipment. However, a spokesman for the Russian Ministry of Foreign Economic Relations subsequently stated that the estimates given by Aven for 1991 and 1992 had no official status.[12] In correspondence with SIPRI Nikolai Revenko, the

[8] Vorobyev, A., 'Interview with G. G. Yanpolskiy, general director of the Ministry of Industry Defense Industry Department', *Krasnaya Zvezda*, 29 Aug. 1992, pp. 1 and 2, in FBIS-SOV-92-173, 4 Sep. 1992, p. 23.

[9] Quotation taken from a summary of the Ostankino (Russian Television) broadcast presentation, 'Russia and arms sales', 10 Aug. 1993, in FBIS-SOV-93-154, 12 Aug. 1993, p. 11.

[10] *Nezavisimaya Gazeta*, 29 Sep. 1992.

[11] Sneider, D., 'Russian armsmakers take off on their own', *Christian Science Monitor*, 25 Nov. 1992, p. 6; *Defense News*, 7–13 Dec. 1992, p. 44.

[12] Correspondence with Peter Litavrin, Department for Export Control and Conversion, Russian Ministry of Foreign Affairs, 12 Jan. 1993.

Counsellor for Disarmament and Control over Military Technologies at the Ministry for Foreign Affairs, confirmed this and repeated that no official information is available after the figure of $1.55 billion given for 1991.

The Chairman of the Russian Committee for Defence Industries Viktor Glukhikh stated that Russian arms exports in 1992 were worth $4 billion.[13] This figure has also been used by Lieutenant General Andrey Nikolayev, First Deputy Chief of Staff of the Supreme Military Command.[14] In November 1993 Glukikh stated that revenue from arms exports would be $3.5–4 billion in 1993.[15]

On 2 December 1992 in a speech to the Russian Supreme Soviet then Prime Minister Yegor Gaidar stated that Russia had concluded agreements worth a total of $2.2 billion in 1992 with three countries—China, India and Iran.[16] In November 1993 Deputy Prime Minister Alexander Shokhin announced that the value of arms exports for 1992 was $2.3 billion.[17]

According to officials of Oboronexport, one of three government-owned export agencies, the value of arms delivered to China alone in 1992 was around $1.8 billion.[18] The Russian State Statistical Committee reported in August 1993 that the value of arms exports for the first half of 1993 was $546.1 million, of which more than 90 per cent went to developing countries (including China).[19]

In December 1993 the Minister for Foreign Economic Relations, Oleg Davydov, released a new estimate that Russia had exported arms worth $1.2 billion in 1993.[20]

Reviewing these statements underlines the current lack of co-ordination between departments of the Russian Government and suggests that there is a high degree of competition between them for

[13] *East Defence & Aerospace Update*, 16–31 Jan. 1993, p. 1; *Frankfurter Allgemeine Zeitung*, 12 Feb. 1993, p. 16.

[14] Moosa, E., 'Russia proposes demilitarized zones in Far East', *Reuters Tokyo*, 24 Feb. 1993.

[15] Statement by Viktor Glukhikh, chairman of the State Committee for Defense Industries, in Moscow Russian Television Network broadcast, 30 Nov. 1993, in FBIS-SOV-93-228, 30 Nov. 1993, p. 49.

[16] The deal with China accounted for $1 billion of this, India and Iran $650 million and $600 million respectively, *Defense News*, 7–13 Dec. 1992, p. 3.

[17] *Süddeutsche Zeitung*, 2 Dec. 1993, p. 11.

[18] *East Defence & Aerospace Update*, Oct. 1993, p. 4.

[19] *East Defence & Aerospace Update*, Aug. 1993, p. 6. Interestingly, the committee included Yugoslavia on the list of recipients for Russian arms.

[20] Quoted by Erik Whitlock in Bergstrand, B.-G. *et al.*, 'World military expenditures', SIPRI, *SIPRI Yearbook 1994* (Oxford University Press: Oxford, 1994, forthcoming).

Table 6.1. Arms exports by Czechoslovakia, 1987–91

Figures are in CSK b. at current prices.

	1987	1988	1989	1990	1991
Total	22 740	19 068	12 195	7 907	5 173
Of which					
Former Socialist countries	17 055	15 134	11 179	6 305	1 581
Other countries	5 685	3 934	1 016	1 602	3 592

Source: *Defence Conversion and Armament Production in the Czech and Slovak Federal Republic*, Background paper submitted to the NATO–CEE Defence Conversion Seminar, Brussels, 20–22 May 1992.

competence in this area. The data released by the Ministry of Industry, Ministry for Foreign Economic Relations, State Committee on the Defence Industries, the office of the Chief of Staff and the office of the Prime Minister are all contradictory. Only the Ministry for Foreign Economic Relations provided any supporting documentation or clarification of the figures released.

Central European countries

Some official data on the arms exports of Central European countries have also been made available in recent years. In several cases this information was produced in the context of NACC discussions of conversion, notably at a seminar held in Brussels in May 1992.

Available data indicate the declining value of arms exports from Czechoslovakia. Moreover, because the data are not adjusted for inflation, the real decline has been even greater than indicated in table 6.1. The decline is the result of the collapse of the trade within the WTO.

In the case of Poland the picture is not as consistent. While table 6.2 shows a dramatic fall in the value recorded for exports between 1987 and 1990, the data for 1991 show a significant recovery. However, the trends indicated differ according to whether they are measured in US dollars or roubles. It is not clear how the exchange rate was calculated for any of the years. Moreover, as all the data are expressed in current prices they reflect the impact of the rapid inflation experienced by

Table 6.2. Arms exports by Poland, 1987–92

Figures are in current prices.

	1987	1988	1989	1990	1991	1992
US dollars	274.4	258.2	188.3	64.9	396.2	67.3
Roubles	1 131.3	5 279.7	992.5	768.2		

Source: For the years 1987 to 1989 the data are contained in Zukrowska, K., *Organisation of Arms Exports in Eastern Europe and Prospects for Limitation of Arms Transfers*, PISM Occasional Paper, Warsaw 1990. For the years 1991 and 1992, the data was provided by the National Statistical Office, Warsaw.

Table 6.3. Arms exports by Hungary, 1971–89

Figures are in percentages.

	1971–75	1976–80	1981–85	1986–89
Domestic sales	52.5	54.2	32.6	20.0
Rouble exports	42.1	36.7	55.1	64.4
Non-rouble exports	5.4	9.1	12.3	15.5

Source: *Defence Conversion and Economic Transformation in Hungary*, Background paper submitted to the NATO–CEE Defence Conversion Seminar, Brussels, 20–22 May 1992.

Poland in recent years. Consequently, as trend indicators the data are not very helpful.

In the case of Hungary there are no official data available on the value of arms exports. However, table 6.3 indicates that the Hungarian defence industry became progressively more export-dependent during the 1980s with other WTO members absorbing a growing share of production. For this reason the collapse of the trading system within the WTO group must have had a particularly severe impact on the Hungarian defence industry.

III. The volume of exports from Central and Eastern Europe

Until recently there were also no official aggregate data describing the number of weapon systems exported by the countries of Central and

Table 6.4. Exports reported in 1993 to the UN Register of Conventional Arms for arms transfers in 1992 by Central and East European countries

Exporting Country	Category[a]	Importing country	No. of items	Description/ comments
Belarus	Tank	North Korea	19	..
Belarus	Tank	Oman	5	..
Bulgaria	LCA	Syria	210	..
Bulgaria	Cbt Acft	Russia	3	..
Czech Rep.	LCA	Zimbabwe	20	Type RM-70 122-mm rocket launcher
Poland	ACV	Latvia	2	..
Romania	LCA	Cameroon	12	130-mm gun
Romania	LCA	Moldova	51	Amphibious armoured carrier
Romania	LCA	Moldova	30	120-mm rocket launcher
Romania	LCA	Moldova	18	122-mm howitzer
Romania	LCA	Nigeria	5	122-mm/40 MLRS
Romania	LCA	Nigeria	4	130-mm gun
Russia	Tank	Oman	6	..
Russia	Tank	UK	1	..
Russia	ACV	Finland	84	..
Russia	ACV	Sierra Leone	4	..
Russia	ACV	UAE	80	..
Russia	ACV	Uzbekistan	30	..
Russia	Cbt Acft	China	20	..
Russia	Cbt Acft	China	6	Training aircraft
Russia	Warship	Iran	1	..
Russia	Warship	Finland	1	Leased unarmed as a museum piece
Russia	Warship	Poland	3	Payment for warship leased to Poland in 1991
Russia	M/Ml	China	144	..
Slovakia	Tank	Syria	81	T-72

[a] *Abbreviations:* Tank: main battle tank; LCA: large calibre artillery; ACV: armoured combat vehicles; Ship: warships; Cbt Acft: combat aircraft; M/Ml: missiles and missile launchers.

Eastern Europe. With the first year of reporting to the United Nations Register of Conventional Arms information of this kind is now becoming available (see tables 6.4 and 6.5).

Table 6.5. Imports reported in 1993 to the UN Register of Conventional Arms for arms transfers in 1992 by Central and East European countries

Importing Country	Category[a]	Exporting country	No. of items	Description/ comments
Bulgaria	Cbt Acft	Russia	5	..
Lithuania	ACV	Russia	15	Type BTR-60 PA
Lithuania	Ship	Russia	2	Light frigate Project-1124
Poland	Ship	Russia	3	Payment for previously leased warships
Romania	Cbt Acft	Moldova	1	MiG-29 fighters

[a] *Abbreviations*: ACV: armoured combat vehicles; Ship: warships; Cbt Acft: combat aircraft.

Table 6.4 contains the data submitted by Central and East European countries in their export returns made to the UN Register of Conventional Arms. Table 6.5 is derived from the reports submitted by Central and East European countries about their arms imports in 1992. Looking at tables 6.4 and 6.5 together it can be seen that some discrepancies appear between the data they contain. For example, Bulgaria reports receiving 5 combat aircraft from Russia while Russia reports no exports to Bulgaria. Similarly, Lithuania reported imports of both BTR-60 armoured vehicles and warships from Russia while neither country was listed as a recipient by Russia in its export return. Discrepancies of this kind were by no means unusual during the first year of reporting to the UN Register.

The UN Register may emerge in time as a major new resource which helps in gaining a comprehensive understanding of the flow of major weapons. It is occasionally supplemented by official data of other types released into the public domain.

Russia

In the context of the discussion of arms transfers to the Middle East between the five permanent members (P5) of the UN Security Council in 1991 and 1992, the P5 governments exchanged some data

Table 6.6. Regional distribution of deliveries of arms and military equipment by the former Soviet Union in 1991

Recipient region	Percentage
Near East	8
Middle East	61
Europe	12
Africa	1
Latin America	1
Asia	17

Source: *Nezavisimaya Gazeta*, 29 Sept. 1992.

on deliveries of major systems in the year 1991. This was a confidential data exchange. However, the then Soviet Government subsequently took a unilateral decision to release its data set to the public (see tables 6.6 and 6.7).

The information released by the Soviet Government described the regional distribution of deliveries of arms and military equipment by the former Soviet Union in 1991 and the balance between different categories of major systems delivered in that year.

Three countries—China, India and Iran—now dominate the discussion of Russian arms exports. Although Russian officials have held discussions with many countries regarding arms sales, few new agreements have been concluded.[21] Russia has found new customers after 1989—most notably China, Iran and Turkey. Two traditionally important relationships with Cuba and North Korea were resumed in 1992, although only through the provision of spare parts.[22]

[21] New countries where Russia is marketing arms include Argentina, Brazil, Chile, Greece, Indonesia, Malaysia, Oman, the Philippines, South Korea, Turkey, the United Arab Emirates and the United Kingdom. Of these countries only Turkey and the UAE have placed orders for equipment, although Malaysia is likely to do so. Pakistan and Taiwan both denied reports that arms sales are under discussion, although Russia and Taiwan have discussed technical and scientific co-operation in aerospace.
[22] Information from Moscow INTERFAX, 3 Nov. 1992, in FBIS-SOV-92-214, 4 Nov. 1992, p. 14; *Far Eastern Economic Review*, 20 Aug. 1992, p. 7.

Table 6.7. Distribution by weapon category of deliveries of arms and military equipment by the former Soviet Union in 1991

	Number of items
Tanks	553
Armoured combat vehicles	658
Large-calibre artillery	381
Combat aircraft	40
Combat helicopters	1
Surface ships	3
Missiles	1 783
Air defence complexes	1

Source: *Nezavisimaya Gazeta*, 29 Sep. 1992.

Ukraine

The great majority of deliveries came from Russia—where the bulk of existing weapon inventories and arms-production capacity are located.[23] However, some deliveries were made by Ukraine—where most of the non-Russian arms-production capacity of the former Soviet Union was located. According to the Ukrainian National Institute for Strategic Studies the value of transfers from Ukraine in 1992 was $962 million.[24] How this figure was calculated is not clear and it seems unlikely that Ukrainian arms exports are all conducted on a hard-currency basis. Moreover, Ukraine returned a nil report for 1992 to the UN Register of Conventional Arms. One country with which Ukraine has established close contact is Iran which is able to supply oil, a commodity which has been contentious in Russian–Ukrainian relations.[25] Ukrainian industrial and government representatives were active in 1992 in important Russian arms markets such as India. After several rounds of discussions India and Ukraine

---

[23] The percentage of arms production capacity from the former Soviet Union located in Russia is in the region of 65–70%. The lower estimate is by the US Defense Intelligence Agency, the higher by Prof. J. Cooper of the University of Birmingham and B. Horrigan in *RFE/RL Research Report*, 21 Aug. 1992.

[24] Correspondence with Serhiy Pirozhkev, National Institute of Strategic Studies, Kiev.

[25] Whitlock, E., 'Ukrainian–Russian trade: the economics of dependency', *RFE/RL Research Report*, vol. 2, no. 43 (29 Oct. 1993).

concluded a trade deal including military equipment on 17 October 1992.[26]

Ukraine depends on machine-building and metalworking industries that make sub-assemblies for shipment to Russia rather than having an independent capacity for system integration.[27] Not only has Ukraine lost much of its traditional market, but the nature of its industrial activity further complicates the formation of new relationships. Bilateral agreements within the CIS should in theory have allowed continuity in inter-republican trade. However, the breakdown of the administrative system of the former Soviet Union has meant that few agreements have been implemented.

The Czech Republic

While most international attention has focused on the Slovakian arms industry it seems likely that the Czech Republic will play a more important role in the international arms market.

As noted earlier, under the new conditions in the Czech Republic sales are more profitable per unit than previously especially if an export customer can be found. However, whereas under the previous system Czechoslovakia delivered more than 200 aircraft per year to other members of the WTO overseas, orders currently stand at around 130 units in total of which 30 are to be supplied from stock—aircraft built as part of an order for the Soviet Union which was cancelled. With a reduction in production volume of this scale cut-backs in employment in the military part of the aircraft industry are inevitable.

In addition to the demonstrated success achieved by Aero Vodochody in selling its L-39 series of aircraft, the Czech Republic has also developed a range of other products which may be marketable overseas. Recently one of these products—the Tamara, a passive sensor capable of detecting aircraft—drew attention to the expertise of the Tesla plant in Pardubice which manufactures a range of electronic systems. Meanwhile, military trucks produced by Tatra might also find foreign customers.

[26] *Asia–Pacific Defence Reporter*, June–July 1992, p. 25; *RFE/RL Research Report*, vol. 1, no. 43 (30 Oct. 1992), p. 61. The deal is to be financed in part through the barter of Indian consumer goods and in part in hard currency.

[27] *Quarterly Economic Review of the EBRD*, 30 Sep. 1992, p. 70.

The Slovak Republic

Prior to the collapse of the WTO important production lines for land systems—artillery and main battle tanks—operated in Czecho-slovakia. Many of the export orders agreed by the Soviet Union with developing countries were actually met with production from these lines.[28]

In 1991 it initially appeared as if the Slovak region would find international customers for its products. An important agreement was reached with Syria for the provision of T-72 tanks with the possibility of follow-on orders from the same country. This agreement was financed by Saudi Arabia as compensation for Syria's role in the 1991 Persian Gulf War. However, after independence the Slovak Republic has not been able to continue with its exports, and the agreement with Syria for a follow-on order for T-72 tanks was cancelled when ZTS Martin—the facility responsible for tank production—was unable to sustain its relations with subsystem suppliers in the Czech Republic.[29]

The range of armoured personnel carriers co-developed by Czecho-slovakia and Poland but produced in Slovakia are also no longer in production.

Poland

Like Czechoslovakia, Poland maintained production lines which were an important element of WTO exports. However, as indicated in table 6.1, deliveries of major conventional weapon systems from Poland had effectively ceased by 1992. There is no evidence that any new sales were concluded in 1993.

As discussed in chapter 4, in Poland arms production has been consolidated into 47 enterprises which are licensed to make and sell military equipment. Of these 30–35 represent a core defence sector where the bulk of arms production occurs. This is a considerable rationalization compared with more than 80 enterprises employing 260 000 people engaged in 'special production' in 1991–92.[30] How-

[28] Cutler, R. M., Després, L. and Karp, A., 'The political economy of East–South military transfers', *International Studies Quarterly*, vol. 31, no. 3 (Sep. 1987), pp. 273–99.

[29] McNally, B., 'Slovakian firm cancels T-72 tank contract with Syria', *Defense News*, 2–8 Aug. 1993, p. 8; *MEDINA Newsletter*, vol. 1, no. 13 (16 Aug. 1993), p. 2; *Intelligence Newsletter*, 16 Sep. 1993, p. 7.

[30] Mesjasz, C., *Problems of Conversion in Central–Eastern Europe: The Case of Poland*, Working Paper no. 15 (Centre for Peace and Conflict Research: Copenhagen, Sep. 1992).

ever, the core of the Polish arms industry always appears to have been 40–45 enterprises with the rest making a small volume of civil items used by the military or dual-use items.[31]

This core defence industry in Poland seems likely to be sustained almost entirely by government subsidy or orders for the national armed forces. There is no strong evidence of sales in international markets. Recent attention has focused on the possibility of selling 320 main battle tanks—a Polish version of the Russian T-72 tank—to Pakistan. According to Jan Straus of the Ministry of Foreign Economic Co-operation, Poland granted a licence for this sale and an advanced stage was reached in discussions with the former Pakistani Chief of Army Staff General Asif Nawaz when he visited Poland in December 1992.[32] However, this contract was never awarded.

Romania

One interesting piece of information revealed in the first year of the UN Register of Conventional Arms was that Romania is an international supplier of 122-mm calibre rocket artillery as well as 122-mm and 130-mm calibre tube artillery. However, the total number of systems transferred is small and the most important single client—Moldova—seems likely to be receiving the systems as aid rather than for cash payment.

IV. Concluding remarks

From these brief descriptions of the prospects of exports of major systems by the countries of Central and Eastern Europe together with descriptions of the recent trends in arms exports it is possible to summarize the prospects for regional exports.

Russia has been able to find a small number of important foreign clients with which it is likely to be able to continue doing business. The economic benefits from these transfers are unlikely to be great and certainly not as sizeable as is hoped by some in Russia. Nevertheless, the agreements already reached and others which can

---

[31] Cupitt, R. T., 'The political economy of arms exports in post-communist societies: the cases of Poland and the CSFR', *Communist and Post-Communist Studies*, vol. 26, no. 1 (Mar. 1993), pp. 87–103.

[32] Jan Straus, quoted in *IDSA News Review–South Asia*, vol. 26 no. 2 (Feb. 1993), pp. 75–76.

realistically be anticipated will be sufficient to make sure that Russia remains an important actor in the international arms trade.

In Central Europe it seems very unlikely that any of the countries will continue as successful exporters with the exception of the Czech Republic. The Czech Republic will, for the foreseeable future, be able to win a share of the international market for jet trainer aircraft and may also be able to market versions of the L-39 series successfully as lightweight fighter aircraft. While the occasional success in export markets is possible, no other country in Central Europe appears to have major systems available for export which can reasonably hope to succeed in an international competition.

As described in chapter 5, there was a degree of integration among the defence industries of Central European countries although in each case this was less significant than the bilateral relationship with Moscow. Consequently, the extent to which Central European countries remain active as arms exporters will also depend on the degree to which they restore defence-industrial ties between them.

Finally, all of the Central and East European countries are producers of light weapons and ordnance. This type of production is not considered at all in this chapter. Information on the trade in these systems is scarce and unreliable while the industrial and strategic importance of the systems is small.

7. Conclusions

Ian Anthony

I. The defence industries of Central and Eastern Europe in context

The general perception of defence industrialists in Central and Eastern Europe is that defence industries are giants, dominating the national economies. In this regard, current perceptions are hostages of the 'worst case' projections of WTO defence industrial capacities developed during the cold war on the basis of fragmentary information.

The findings in this report suggest that—apart from Russia— defence industries represent only a small proportion of national industrial capacity in Central and Eastern Europe. Moreover, while Russia undoubtedly has a major defence industry in international comparative terms, even here the dominance of this sector within the economy has been exaggerated. The defence industry is a relatively small employer in the Czech Republic, Hungary and Romania. Even in Poland and Slovakia—whose industrial sectors are widely supposed to be defence-dependent—there is insufficient evidence to support this conclusion.

There are two likely explanations for the above misperception. First, there is the lack of adequate definitions to describe the defence industry. This is primarily an academic failure. Second, there is the lack of information about the nature and scale of defence industrial activities in Central and Eastern Europe. This is a failure of governments in the region. Clearly, students of the defence industry, both within and outside government, have something to teach but much more to learn.

The future size of the defence industries in Central and Eastern Europe is impossible to predict with any precision. None of these countries has yet determined the size, structure or operational doctrine of its armed forces. These decisions will ultimately determine the level of stable demand for military equipment and, therefore, condition the size of the industry. However, it is already possible to make some forecasts about the size and shape of future industrial capabilities with a fair degree of confidence for at least some of these countries.

Bulgaria, the Czech Republic, Poland, Romania, Slovakia and Ukraine all have defence industries which produce a limited range of products. Moreover, they have little or no indigenous military research and development. Therefore, the current range of products is not likely to increase by the introduction of new designs but through incremental modifications to existing designs. This is even more the case for Hungary.

In Europe, the military operational environment is in a process of continuous change as the technological advances of the 1980s are assimilated by NATO member states and, to a lesser extent and at a slower pace, by Russia.[1] Separated from one centre of military technology innovation—Russia—without being linked to the other—the USA—it seems unlikely that smaller defence industries can survive over time. Rather, they will increasingly be manufacturing equipment that is obsolete in the European context. While this equipment might be of military utility in some (although not all) developing countries, this market is too uncertain to sustain significant industrial capacities without permanent government subsidies.

In Russia, the picture is different. The Russian Government maintains that it should keep military capabilities beyond those needed for territorial defence. These capabilities would be needed to defend Russian interests across large areas of Europe and Asia, to employ in peace-enforcement operations on the territory of the former Soviet Union and, perhaps, to make a major contribution further afield in the context of United Nations military operations. Moreover, the Russian defence industry is not only large but has demonstrated that it is capable of innovation across the full spectrum of military equipment.

It is theoretically possible for Russia to achieve this level of military capability. However, the Russian defence industry is in such deep crisis that the sustained investment needed to do so seems unlikely to materialize. Lack of investment and, perhaps equally important, the feeling that the sector has no future appear to be 'hollowing out' the defence industry. Technicians and other skilled workers are not being

[1] Analysing the broad impact of this change, which is increasingly being described as the 'military technology revolution', is the subject of the SIPRI Project on Military Technology and International Security. It is supported by the W. Alton Jones Foundation and led by Dr Eric Arnett and Dr Rick Kokoski of SIPRI.

laid off but are voluntarily leaving the sector while young people with a technical education are not attracted to it. If this continues for any length of time (as seems inevitable) then Russia's capacity for innovation in the military area will waste away. The Russian defence industry is so large that even after rationalization and concentration, what remains will still be significant in global comparative terms.

These prognoses are simple extrapolations of current trends. However, the current situation is so fluid that what seems most likely today could change rapidly tomorrow. Therefore, it is possible to imagine developments that might alter this pattern and some of these are considered in the final section.

II. Structural adjustment

It is generally acknowledged that the sale of defence goods has a quite different dynamic from the sale of civil goods because the government dominates all aspects of the defence market. At least one study has argued that defence producers in market and planned economies have more in common with each other than either has in common with civilian manufacturers.[2] Defence contractors have no independent control over which products to develop or the volume of production. In comparison with their counterparts in the civilian markets, defence producers have limited possibilities to shape demand, which is set by the requirements of the armed forces. Neither can they set prices for the products they make independently. Defence producers in market economies no less than in command economies must bargain with government to determine the financial terms for any given programme.

The defence industries of Central and Eastern Europe share some of the fundamental problems which are having an impact on defence industries elsewhere. These problems stem in large part from the unwillingness of governments to allocate the same level of resources to defence after the end of the cold war. The reduced volume of global arms sales (which is itself partly a consequence of the end of the cold war) has also had an impact.

In the face of reduced military expenditure defence industries are under pressure to perform several mutually exclusive tasks. First, they

---

[2] Gorgol, J. F., *The Military Industrial Firm: A Practical Theory and Model* (Praeger: New York, 1972), especially appendix A.

must operate according to commercial principles in a regulatory environment which mandates the retention of unproductive capacities. Second, they must operate at little or no cost to the national treasury in a period of low international demand for military equipment.

While the nature of the problem—reduced military expenditure—is common to many OECD countries, the scale is not. In most OECD countries reductions in procurement spending have been gradual compared with Central and Eastern Europe. Moreover, not only have the reductions in overall military expenditure across Central and Eastern Europe been sizeable, but within these reduced budgets the allocation to procurement of major equipment has been heavily reduced. Procurement accounts for a reduced share of a reduced budget. This picture is blurred to some extent by the fact that the defence industry also receives significant extra-budgetary allocations. However, it appears that these additional monies are being used largely for social welfare purposes rather than to sustain industrial production.

Jacques Gansler offers an evaluation of the US defence industry which has a degree of resonance for the defence industries in Central and Eastern Europe. Gansler observes that the defence industry is a 'captive sector' of the economy, 'dependent on the Department of Defense and largely isolated from the commercial economy by a wall of government regulation and red tape . . . burdened by exorbitant debt, excess production capacity, a rapidly shrinking market, escalating unit costs, lengthening development cycles and a fundamental loss in business confidence.'[3]

While elements of this description could equally be applied to the defence industries of Central and Eastern Europe, there are also some important differences. These differences exist at both the wider societal level and as regards the more specific organization of the business units that make up the defence industry.

No government—including that of the United States—has yet developed a clear policy designed to match its defence industrial capacity to post-cold war realities. However, many if not all OECD countries have demonstrated their ability to absorb reductions in capacity in other industrial sectors which employed large numbers of people and generated a significant percentage of national income—

[3] Gansler, J. S., 'Transforming the US defence industrial base', *Survival*, vol. 35, no. 4 (winter 1993/94), p. 134.

such as the textile, coal, steel or shipbuilding industries. The defence industries of the member states of the OECD are undoubtedly facing local difficulties, and certain regions will take time to recover from the social and economic impact of declining defence industrial activity. Nevertheless, few doubt that this adjustment will be made successfully even in countries such as France, the United Kingdom and the United States, whose manufacturing industries have come to regard themselves as 'defence dependent'.

By contrast, the economies of Central and Eastern Europe have not yet demonstrated their capacity to make adjustments of the kind described above. Moreover, there are structural impediments to adjustment which have not yet been overcome. One of the most serious is the absence of business units comparable to companies which operate in market economies. Structural changes in the organization of business units in Central and Eastern Europe are being undertaken.

Large companies in a market economy combine the tasks of product development, manufacturing and distribution in one company or industrial group. However, these elements of the overall production process have been almost entirely separated in Central and Eastern Europe with government agencies linking them together. Through transfer of management authority to a higher level (often called a joint stock company but playing the role more usually associated with a holding company) these functions are now being aggregated in single business units. In most cases these new business units remain in public ownership either directly or because they are owned by state-controlled banks. Limited private investment in defence industries is permitted provided this does not threaten state control over decision making.

In the early stages of transition away from a command economy the emphasis in decision making was placed on decentralization of authority. However, it is now becoming clear to both industrial and political leaders that breaking up business units is not sufficient. They must be reassembled in a more productive manner.

In the context of the defence industry the word privatization is usually used to describe one of the processes through which this transformation is being attempted. As a result, privatization can mean several things. For example, the word can be used to clarify the extent of state ownership. This clarification is needed to determine who should manage the process and control the proceeds received from the future sale of state assets. Another use is to describe investment of non-state capital (either national or foreign) into defence industry enterprises. This investment is not often allowed to exceed 50 per cent of the value of the defence enterprise and the state retains ultimate control. However, the word 'privatization' may also be used to describe the transfer of ownership from the state to private individuals. This process has hardly begun in the defence sector, although it is more widespread among the non-defence activities of diversified industrial units and among industrial units engaged in dual-use production.

The process of restructuring is not unique to the defence industry but is taking place across the spectrum of manufacturing industry in Central and Eastern Europe. However, whereas in most civilian areas restructuring is seen as a preparatory step for ultimate transfer to private ownership, this is by no means clear in the defence sector. Even in Central European countries such as Hungary, Poland, Romania and Slovakia, the defence industry is likely to remain in public ownership. In Russia and Ukraine, continued government ownership seems almost certain. Therefore, government agencies will play a central role in the process of reassembling industrial assets in the defence sector.

This does not mean that some of the elements currently thought of as part of the defence industry will not eventually become privately owned. As noted in chapter 1, the 'defence industry' is not an industrial sector whose membership has been clearly defined. Many of the entities which previously fell under the administrative control of government or communist party agencies dealing with the defence industry were purely civil in character. It is primarily these activities which are currently being prepared for sale to private shareholders. However, whether or not the business units which conduct these activities reorient their sales to non-government customers is likely to reflect a market-led decision by their new managers rather than an administrative directive.

In the OECD countries, defence industries are adjusting to the loss of sales which they have experienced or anticipate through a variety of steps.[4] Adjustment strategies have included: reducing employment;

[4] This process of adjustment is more fully described elsewhere by SIPRI in chapters on arms production in successive *SIPRI Yearbooks* after 1990 and in two other recent volumes:

seeking subsidies from government; industrial concentration (national and, to a lesser extent, transnational); and forming partnerships and joint ventures with other companies to share financial risks and seek new overseas markets.

As this report underlines, defence producers in all of the Central and Eastern European countries have examined all of these adjustment strategies.

Large and compulsory reductions in employment have been avoided to the end of 1993. While a significant number of people have left the sector, this has largely been a voluntary decision by individuals in major cities where alternative employment is more readily available. In addition to a lack of alternative employment opportunities, the dependence of employees on the workplace to meet many of their social needs has been a major factor deterring large-scale lay-offs. Much anecdotal evidence suggests that many individuals nominally still employed in the defence industry in fact have more than one job.

While the impulse to maintain levels of employment comes from both government and industry, the decision whether or not to continue subsidies is a political choice. Many feel that at some point a 'credit crunch' will have to come. However, the probability is that subsidies and support will first be cut off from the politically weak. Issues of technological competitiveness or profitability are likely to be secondary to the level of political influence any enterprise or sector can wield. In this regard the defence industries in Central and Eastern Europe are better placed than many other sectors.

In Russia, for example, important figures in the defence industry are also significant players in domestic politics. Influence is exercised along three different avenues. The defence industry has moved more quickly than some others in setting up sectoral industrial associations to represent their interests and lobby both executive and legislative branches of government. Second, many of the most powerful executive bodies established by President Boris Yeltsin have at their core individuals with close ties in the defence industry. Finally, individuals with close ties to the defence industry were among those elected to the new Russian Parliament in December 1993.

Brzoska, M. and Lock, P. (eds), *Restructuring of Arms Production in Western Europe* (Oxford University Press: Oxford, 1992); and Wulf, H. (ed.), *Arms Industry Limited* (Oxford University Press: Oxford, 1993).

These observations need to be kept in perspective, however. Attempts by industrial lobbies to exert political influence are not unusual and there is no evidence that the defence sector is unique in this regard. For example, equivalent efforts by industrialists from the oil and gas industry and the nuclear energy sector are under way. Moreover, defence industrialists are not monolithic in their political views. One observer has noted that there are at least three groups of industrial managers with different political inclinations.[5]

One group is interested in law, order and stability which they need in order to be able to operate their businesses successfully. Although their interests are not primarily partisan, many of this group are associated with Russia's Choice and the Civic Union. A second group would like to see the restoration of the old political order or something closely resembling it. A third group—by far the largest—consists of managers who recognize the uniqueness of the current conditions and are learning as much as they can as fast as they can about survival in a market economy. Within this group many would prefer to leave the defence sector entirely but do not have the necessary capital or conditions.

The defence industries of Central and Eastern Europe have also sought increased foreign contact as a possible solution to their problems. Both industrial joint ventures with foreign partners and exports of defence equipment have been sought. However, neither approach has led to significant successes yet.

International joint ventures have been formed for two different purposes which must be evaluated separately. First, there are the joint ventures in the defence sector and, second, the joint ventures as a form of diversification away from defence.

In the short term commercially based industrial joint ventures in the defence sector are most likely where they are based on existing expertise in carrying out the repair, maintenance and upgrade of equipment of Soviet origin. If in future the procurement budgets of Central and East European countries begin to grow, joint ventures may be a means of gaining new market access for foreign companies. Joint development of new systems seems very unlikely to occur.

---

[5] Jevgeniy Kuznetsov, Centre for Economic and Mathematical Studies, Moscow, at the Försvarets Forskningsanstalt (Swedish National Defence Research Establishment, FOA) seminar on the Future of the Russian Defence Industry, Stockholm, 21–22 Oct. 1993.

The issue of joint ventures is also likely to emerge in the framework of joint operations proposed in the framework of the NATO Partnership for Peace programme. Central European countries face barriers to foreign co-operation in that the country with which they have the most in common in terms of technology and equipment—Russia—is exactly the country on which they are seeking to reduce their dependence. In joint operations, questions of equipment standardization will be difficult to overcome in the absence of large-scale military assistance from NATO, which seems unlikely to be forthcoming.

Under these circumstances it would be logical for Central European countries to seek closer industrial ties with one another—something long proposed by industrialists themselves. However, there are political obstacles to this approach.

International joint ventures as a form of diversification seem unlikely to be created in any significant number. Defence producers make unattractive partners for foreign investors, most of whom are interested in market access. Therefore, they are more likely to seek partnerships with civil industries with a proven distribution system.

Increasing foreign sales has also been examined as a partial solution to the problems facing the defence industries of Central and Eastern Europe. However, no evidence exists to indicate that a strategy based on foreign sales can succeed. By the end of 1993, Central European countries had failed to sell more than a handful of systems abroad, and only Russia had succeeded in capturing a potentially lucrative new market (in China). While Russia is likely to remain an important player in the international arms market, none of the other countries of Central and Eastern Europe seems poised to do so, with the exception of sales of specific products such as Czech jet trainer aircraft.

In foreign markets, producers face both political and economic barriers to sales where they have traditionally been successful. Sales to many of these countries are now prohibited either because the recipient is proscribed under new export regulations or because there is concern that sales will provoke a negative reaction in the United States. At the same time, after the cold war the rationale for arms sales is less political and more commercial for countries of Central and Eastern Europe. In some cases the traditional recipients of arms from Central and Eastern Europe are unable to pay for new equipment. Producers in Central and Eastern Europe have few of the possibilities which exist among their competitors to provide financial assistance to potential customers.

In opening new markets, Central and East European industrialists face strong competition from traditional suppliers—mostly from North America and Western Europe—as well as having to overcome the sense among buyers that current political and economic uncertainty in the region makes them unreliable partners.

III. Long-term developments

If it is clear that there will be great short-term difficulties in making structural adjustments, to what extent have long-term decisions that will shape defence industrial policy been made? What kind of defence industry is desired?

In Central Europe many of these decisions have been taken. While none of the countries of Central and Eastern Europe is ready to do without a defence industry, all will adopt a 'minimalist' policy to maintain production of ordnance and spare parts for equipment in service. This involves some limited investment in new production capacities to replace spare parts and perhaps even major sub-assemblies such as engines previously supplied from within the WTO production system. No new products will be developed but a few incremental modifications to existing systems will go ahead—for example, the addition of foreign tank night-vision systems or aircraft radars—where capabilities can be enhanced significantly at relatively low cost.

In Russia, by contrast, no clear pattern of future developments has emerged. As noted above, the government has not yet abandoned its aspirations to a defence industry comparable in size and technical capacity with that of the United States. The most probable engagements that the Russian armed forces will undertake are low-intensity operations on the territory of the former Soviet Union (such as recent operations conducted in Tajikistan). However, while maintaining levels of force needed for such operations, military planners have not abandoned plans to develop the capacity to conduct operations of the type US-led coalition forces undertook against Iraq in 1991.

While these plans—many of which were conceived in the early 1980s—to develop new military capabilities and the requisite new generation of weapons and equipment remain part of Russian policy, pressures will remain to allocate more resources to defence than can be justified by Russia's threat environment. Moreover, developments in Russia will be the decisive factor in determining the defence posture of many countries. Under these conditions not only countries in Central and Eastern Europe but also many others besides will tailor their defence efforts to 'worst-case' scenarios in which Russia takes the place of the Soviet Union as the country regarded by many as the primary threat to international security. The failure to reduce the size of the defence industry and tailor it to the security needs of the modern Russian state would not only handicap Russian economic development but also be a self-defeating security policy.

Appendix. List of participants in the 1993 SIPRI workshop

The individuals listed below participated in the SIPRI Workshop on the Future of the Defence Industries in Central and Eastern Europe held on 29–30 April 1993.

I. Outside participants

Oleg Bogdanov
Chief Designer
Antonov Design Bureau
Kiev
UKRAINE

János Csendes
Head of Export Control
Ministry of International
Economic Relations
Budapest
HUNGARY

Oleg Gapanovich
St Petersburg City Council
Military Industry and
Conversion Commission
St Petersburg
RUSSIA

Dr Judit Kiss
UN University
World Institute for
Development Economic
Research
Geneva
SWITZERLAND

László Kocsis
Director General
Ministry of International
Economic Relations
Budapest
HUNGARY

Juraj Kovácik
Manager, Engine Division
ZTS Martin
Martin
SLOVAKIA

Jan Leijonhjelm
Head of Department
National Defence
Research Establishment
Sundbyberg
SWEDEN

Peter Magvasi
Financial Director
ZTS Martin
Martin
SLOVAKIA

Per Olof Nilsson
National Defence
Research Establishment
Sundbyberg
SWEDEN

Nina Y. Oding
Head of Research
Department
Leontief Centre
St Petersburg
RUSSIA

Prof. Maciej Perczynski
Polish Institute of
International Affairs
Warsaw
POLAND

Åke Petersson
Ministry of Foreign
Affairs
Stockholm
SWEDEN

Maj. Gen. Maftei Rosca
Director General
Ministry of Industry
Bucharest
ROMANIA

Oleg Samoylovich
Deputy Chief Designer
Moscow Aviation
Institute
Moscow
RUSSIA

Prof. Arnold Shlepakov
Economic Research
Institute
Kiev
UKRAINE

Adam Stranák
Vice President, Engineering
Aero Vodochody
Prague
CZECH REPUBLIC

Sture Theolin
Head of Division
War Materiel Inspectorate
Stockholm
SWEDEN

Vadim I. Vlasov
Assistant to the First
Deputy Minister of Defence
of the Russian Federation
Moscow
RUSSIA

Dr Jan Straus
Director, Central
Engineering Board
Ministry of Foreign
Economic Relations
Warsaw
POLAND

II. SIPRI participants

Dr Ian Anthony
Project Leader
Arms Production and Arms
Transfers Project

Shannon Kile
Research Assistant
Russia's Security Agenda
Project

Elisabeth Sköns
Researcher
Arms Production and Arms
Transfers Project

Dr Eric Arnett
Researcher
Military Technology
and International
Security Project

**Evamaria Loose-
Weintraub**
Research Assistant
Military Expenditure
Project

Dr Daniel Tarschys
Chairman of the SIPRI
Governing Board and
Chairman of the Committee
for Foreign Affairs of the
Swedish Parliament

Gerd Hagmeyer-Gaverus
Researcher
Arms Production and Arms
Transfers Project

Dr Adam Daniel Rotfeld
Director
SIPRI

Siemon Wezeman
Research Assistant
Arms Production and Arms
Transfers Project

Index

Adams, Gordon 5
Aeroflot 95
Aero Vodochody 48, 88, 89, 102, 109, 119
Afghanistan 107, 108
air defence 27, 31, 100
Alferov, Vladimir 9, 76
Allied Signal 102
Almaz association 90
America *see* United States of America
Angola 107
Antonov 56, 57, 84, 95
arms trade:
  cold war and 107, 108
  credit arrangements 109–10
  data on 108–14
  financial transfers and 109
  income from 108–109
  production and 1–2
  regulation 76–82, 131
artillery 121
Åslund, Anders 7
Australia Group 76
Aven, Peter 111
Aviaexport 56

Barchuk, Vasiliy 41, 42
Belarus 105, 115
Bogdanov, Oleg 95
Bulgaria:
  arms exports 115
  arms imports 116
  CFE ceilings 33, 36
  debts 53
  defence industry 2, 124
  economy 53

military expenditure 53–54
  R&D 3
Butler, William 84

Cambodia 107
Catherwood, Sir Fred 98
Central Europe:
  aid to 98
  armed forces 10, 26–27, 30–32
  arms exports 76–78, 113–22, 131, 132
  arms imports 100
  arms production 67
  financial system 2
  investment, foreign 93
  military doctrine 16–17
  military expenditure 4, 32, 126
  military planning 10, 16
  privatization 71, 73–74
CFE (Conventional Armed Forces in Europe) Treaty (1990) 1, 17fn., 32, 33, 103
China:
  arms imports 108, 110, 112, 117
  Russia and 4, 23, 108, 112
CIS (Commonwealth of Independent States) 40, 48, 94, 99
CMEA (Council for Mutual Economic Assistance) 37, 109, 110
COCOM (Co-ordinating Committee on Multilateral Export Controls) 76, 80, 94, 100
Cooper, Julian 2
CSCE (Conference on Security and Co-operation in Europe) 39
Cuba 107, 117

Czechoslovakia:
  arms exports 113
  arms imports 108
  defence industry 4:
  division of 48
  military expenditure 49
  see also following entry and
  Slovakia
Czech Republic:
  armed forces, restructuring 30–32
  arms exports 80, 115, 119, 122
  CFE ceilings 33, 36
  defence industry 2, 88, 104, 108,
    124
  industry, restructuring 8
  military doctrine 16, 25
  military expenditure 49

Davydov, Oleg 112
defence industries:
  adjustment strategies 128–29
  business units 82, 84–87
  conversion 83
  co-operation, cross-border 91–106
  defining 12, 123, 128
  employment and 1
  enterprises 82
  exports and 1
  foreign suppliers 99
  government regulation 74–76
  industrial restructuring 8
  internationalization 91–106
  intra-enterprise relations 82–90
  investment in, foreign 98
  joint arrangements 4–5, 93–99
  market economies and 82, 83
  marketing 91
  military expenditure and 125–26
  modifications 101–3
  national orientation 91
  over-capacity 1, 99
  ownership 70–74, 84, 128

privatization 71, 127–28
production, maintaining 1–2
production associations 84
restructuring 58, 91–106
SIPRI workshop on 5–6
state ownership 84
units of 13, 14
defence sufficiency 26–27

Eastern Europe:
  aid to 98
  arms exports 114–21, 131, 132
  business units 84, 87, 127
  design organizations 84
  economics in 6
  financial system 2
  investment, foreign 93
  military expenditure, fall in 126
  privatization 71
Egypt 102, 107fn., 108, 109
Elop 101
Europe *see* Central Europe; Eastern
  Europe
European Union 104, 105

France 71, 100, 102, 103

Gaidar, Yegor 47, 112
German Democratic Republic 3, 92,
  108
Germany 81, 94
Glukhikh, Andrey 112
Gorbachev, President Mikhail 28,
  79
Grachev, Pavel 11, 22, 23, 36, 64
Great Britain *see* United Kingdom

Hooks, Gregory 59
Hughes Electronics 99
Hungary:
  armed forces restructuring 30–32
  arms exports 76–77, 81, 114

arms imports 100
CFE ceilings 33, 56
debts 51
defence industry 2, 67, 68, 73–74,
   104, 123, 128
economy 51
industry restructuring 8
military doctrine 16, 22, 25
military expenditure 39, 51–53
NATO and 105
privatization 73–74
procurement 67
R&D 3
Russia and 52, 110

IMF (International Monetary Fund)
   38
India 102, 108–10, 112, 117–19
inflation 37
Iran 112, 117, 118
Iraq 54, 107, 108, 109
Israel 71, 101, 102, 108
Israel Aircraft Industries 102
Italy 71, 94

Kamov 95
Kiss, Judit 2–3
Klimov 95, 99
Kokoshin, Andrey 11, 42, 109
Korea, North 107, 117
Kortunov, Sergey 7
Kuchma, Leonid 73

Libya 107, 109
Lithuania 116
Lyulka 95

Malei, Mikhail 61
market economies 82, 83, 127
Mikoyan 85, 95, 99, 101
military doctrine:
   changes in 16, 18, 93

nature of 18–19
survey of 25–30
military expenditure:
   accounting, variations in 38
   armed forces role and 37
   budgets, exceeding 40
   data on 38–40
   definition 38
   fall in 4, 37–57, 125–26
   market economies and 39
   price distortions 39–40
   reductions 4, 37–57, 125–26
   transition in 37–57
Moldova 121
MTCR (Missile Technology Control
   Regime) 76, 80

NACC (North Atlantic Co-operation
   Council) 39
Nachaev, Andrey 41
NATO (North Atlantic Treaty
   Organization):
   Central Europe and 5, 104
   defence industry collaboration
      91–92
   military expenditure 38
   Partnership for Peace and 21, 131
   WTO and 20
Nawaz, Asif 121
Nicaragua 107
Nikolayev, Andrey 112
NPT (1968 Non-Proliferation
   Treaty) 22, 23
Nuclear Supplier Group 76

Oboronexport 80, 90, 112
OECD (Organisation for Economic
   Co-operation and Development)
   93, 126, 128
oil 118, 130

Pakistan 121

Persian Gulf War 30, 120
Poland:
  armed forces 26, 31–32
  arms exports 77, 81–82, 113–14,
    115, 120–21
  arms imports 100, 108, 116
  CENZIN 77, 81
  CFE ceilings 33, 36
  debts 50, 56
  defence industry 2, 4, 88, 100–101,
    104, 108, 123, 124, 128
  economy 51
  industry, restructuring 8
  military doctrine 16, 22, 25
  military expenditure 39, 50–51
  privatization, meaning of 71, 127–
    28
  production associations 84, 86, 87

Radetskyy, Vitaliy 34
Reagan, President Ronald 70
Rediffusion 99
Revenko, Nikolai 111–12
Romania:
  arms exports 115, 121
  arms imports 102, 116
  CFE ceilings 33, 36
  defence industry 2, 67, 68–69, 73,
    108, 123, 124
  military expenditure 54, 55, 69
  missile production 92
  procurement 67
Russia:
  aircraft production 95–98
  air-mobile units 28–29
  armed forces 11, 17, 24, 35–36
  arms exports 27, 34, 57, 108, 115,
    111–17, 121–22, 131:
  arms imports 99–100
  arms production 43, 46, 84, 85, 86,
    87, 118fn.
  as threat 133

Bashkortostan 66
Basic Law on Defence 64
borders 23, 29, 35
businesses 8
cargo traffic 95
centre–periphery relations 65–67
CFE ceilings 33, 35
China and 4, 23, 108, 112
Civic Union 130
civil aircraft 95–97
civil defence 43
civil sector, developing 8
conflict resolution and 107
conflict within 24
Conversion, Law on 67, 75
Conversion Fund 47
Council of Ministers 61
debts, internal 46, 56, 88
Defence Branches of Industry,
    Committee for 61
Defence Orders 42, 47, 55, 64.
Deliveries of Products and Goods
    for the State, Law on 70
designers in 3–4
design organizations 84, 95
devolution in 65
economic changes 7, 8, 10
electricity generation 95
financial industrial groups 89–90
foreign policy 11, 17
freight transport 95
geostrategic position 28, 34
Government–industry relations
69–   82
Hungary and 52
imperialism, fear of 21, 24
Industrial Policy, Committee for
    61
industry: defency industry and 65
Islam and 23
instability in 21
investment, foreign 98

joint ventures 94, 95–98
Middle East and 111
military aid 107
military districts 35
military doctrine 11, 17, 22, 27–30
military expenditure 39, 40–47:
military operations 24, 124, 132
mobile forces 29
Mobile Forces Command 35
mobilization capacities 74–78
mobilization reserves 28, 29
'near abroad' 24, 124
neighbours 11
nuclear forces 23, 64
Parliament 41, 44, 129
prices in 42
privatization 71–73
procurement 43, 44, 64
Promexport 80
R&D 2–3, 42, 47, 70, 84, 86
Russians abroad 24
State Defence Committee 75
State Defence Orders, law on 70
Supreme Soviet 10
Tatarstan 66
tax 55
technological advances and 124
threat perception 11, 22–24
troop demobilization 34, 40
Udmurtiya 66
Ukraine and 4, 22, 24–25, 118
UN and 124
USA and 107

Sapir, Jacques 87
Saudi Arabia 120
Scud missiles 30
Serbia 81
shipbuilding 43
Shokhin, Alexander 112
Slovakia:
  armed forces restructuring 30–32

arms exports 82, 115, 120
CFE ceilings 33, 36
defence industry 2, 8, 88, 104, 123
investment in 48
military doctrine 16, 25
military expenditure 50
OMNIPOL 82
Russian debt to 110
unemployment 48
see also Czech Republic
SNECMA 99
Sobchak, Anatoliy 65
socialism 82, 84
SOFMA 103
South Africa 71
Spetsvneshtekhnika 80, 90
Straus, Jan 121
submarines 43
Sukhoi 85, 95, 101
Sweden 100
Syria 107, 108, 109, 120

Tajikistan 132
Tamara sensor 119
Tashkent agreement 35
Tatra 48, 119
Tesla plant 119
Thailand 109
Thomson CSF 102
threat perception 19–25
Tupolev 95
Turkey 108, 117

Ukraine:
  aircraft production 2, 90
  armed forces 17–18, 33–34
  arms exports 118–19
  arms production 118–19
  CFE and 33, 36, 48
  defence industry 2, 88, 90, 104,
    105, 124, 128
  equipment inherited 32

ICBM production 2
military doctrine 17–18
military expenditure 39, 47–48
nuclear weapons 18, 23
privatization 71, 73
R&D 2
Russia and 4, 22, 24–25, 90, 118
threat perception 24–25
troops returning 48
unemployment 37
Union of Soviet Socialist Republics:
  arms exports 78, 101, 108–10,
    117–18
  arms production 118
  barter trade and 110
  defence industry 85, 86
  GOSPLAN 9
  military doctrine 18–19
  military expenditure 40
United Kingdom:
  defence industrial base 13
  defence industry 71
  joint ventures 94
United Nations:
  arms exports and 115, 116–17
  embargoes 54, 81, 107
  Military Expenditure, Project on
    38
  Register of Conventional Arms
    115, 116, 118
United States of America:
  conflict resolution and 107
  defence industrial base 12
  defence industry 70–71, 85, 126
  joint ventures 94
  military aid 107
  R&D 85
  Russia and 107

Viet Nam 107
Visegrad Group:
  armed forces needed 25

conflict scenarios 21
defence industry co-operation
  103–106
defence sufficiency 26–27
military doctrines 22, 25–27
NATO and 21–22
threat perception 21–22
Vlasov, Vladim I. 60
Vorobev, Vasiliy 41, 42

Weidenbaum, Murray 70
World War II 74
WTO (Warsaw Treaty
  Organization):
  collapse of 1, 4, 16, 21, 93, 107
  defence industry, integration of 4,
    58, 93
  division of labour in 92–93
  ideology and 20
  'security vacuum' left by 21
  USSR's domination 20, 58, 92

Yakovlev 95
Yanpolsky, G. 111
Yeltsin, President Boris 7, 9, 11, 46,
  58, 75, 78, 80, 129
Yugoslavia:
  airborne surveillance of 100
  embargoes and 54, 81, 107